KT-567-337

Richard [   ] is a senior feature writer for the *Sunday Times M* [   ] He was named Specialist Writer of the Year in the U[   ] Awards in 2002, and shortlisted in 2005. He is the au[   ] *elfstan's Place*, a collection of hi torical s[   ] stories, *Spr g's War*, a novel of the English Civ War, a[   ] *View from the Top*, an illustrated guide to the structure and s-tory of the British landscape. He has been a consultant to the former Department of the Environment and the Department of Culture, Media and Sport, and author of campaigns for the Campaign to Protect Rural England (CPRE). He is currently a trustee of the Tree Council. Richard Girling lives and works in North Norfolk.

'I believe this is one of the most important books of the year. It is also highly fascinating, totally absorbing, compulsively readable and should be pushed into everybody's hands. Read it and be converted.'
Sarah Broadhurst in the *Bookseller*, 25 February 2005

Also by Richard Girling

Fiction
Ielfstan's Place (The Forest on the Hill)
Sprigg's War

Non-fiction
The View from the Top

RUBBISH!

Dirt on Our Hands and Crisis Ahead

Richard Girling

eden project books

TRANSWORLD PUBLISHERS
61–63 Uxbridge Road, London W5 5SA
a division of The Random House Group Ltd

RANDOM HOUSE AUSTRALIA (PTY) LTD
20 Alfred Street, Milsons Point, Sydney,
New South Wales 2061, Australia

RANDOM HOUSE NEW ZEALAND LTD
18 Poland Road, Glenfield, Auckland 10, New Zealand

RANDOM HOUSE SOUTH AFRICA (PTY) LTD
Endulini, 5a Jubilee Road, Parktown 2193, South Africa

Published 2005 by Eden Project Books
a division of Transworld Publishers

Typeset in 11/14pt Times by
Falcon Oast Graphic Art Ltd

Printed in Great Britain by
Bookmarque Ltd, Croydon, Surrey

1 3 5 7 9 10 8 6 4 2

Papers used by Eden Project Books are natural, recyclable products made
from wood grown in sustainable forests. The manufacturing processes
conform to the environmental regulations of the country of origin.

For Caroline

CONTENTS

Acknowledgements

The usual form with acknowledgements is to thank everyone else for their help and encouragements, and to accept personal blame for all mistakes. This is honourable but disingenuous. Help I had in plenty, and my gratitude to those who gave it remains heartfelt. But there were hindrances too. A promised official briefing from Defra never materialized, so it is difficult to accept responsibility for any failure to perceive the cogency of that department's policies. And of course, in the nature of politics, it is the straightest questions that receive the most serpentine answers.

Even in departments of government, however, there are exceptions. Susanne Baker and Matt Conway at Defra were models of helpfulness, and neither is to blame for the use I made of the material they provided. Others to whom I am indebted, and who may also be absolved of blame, include Peter Jones of Biffa Waste Services, Phil Burston of RSPB, Barrie Clarke and Jacob Tompkins of Water UK, Chris Shipway of Thames Water, Tony Harrington of Yorkshire Water, Emer O'Connell and Andrew Skinner of the Environment Agency, Paul O'Grady, Peter Gerstrom, Peter Braithwaite and Mike Chrimes of the Institution of Civil Engineers, Alistair Gammie of Bayer Diagnostics, Nick

Reiter of the Deer Commission of Scotland, Randolph Hodgson of Neal's Yard Dairy, Mark Strutt of Greenpeace, Clare Wilton of Friends of the Earth, Steve Jenkins of Norfolk Environmental Waste Services, Sue Reid of Daventry District Council, Ray Georgeson and Pat Jennings of WRAP, Mark Wheeler of the Health and Safety Executive, Alan Hamilton of *The Times*, Tim Ambler of the London Business School, and Sylvia Wilson, an authentic local hero in Nelson.

To these must be added numerous others who helped with earlier articles for the *Sunday Times Magazine* on which the text also draws, and Robin Morgan, the magazine's editor, for allowing extensive quotation from published work. No *Sunday Times* article is here reproduced in full, but I apologize to any reader who may come across a familiar paragraph or several. Special thanks are due to my editors, Mike Petty at the Eden Project (whose idea the whole thing was) and Susanna Wadeson at Transworld (who nursed the text), and to my hawk-eyed copy editor Deborah Adams (who spared me embarrassment).

It is also usual form to thank one's wife for her encouragement/forbearance etc. Caroline gave me much more than that and I only hope that, by the time these words appear, I shall have thought of an adequate way to thank her.

Richard Girling
Norfolk, 2005

Introduction

To those of us who survive into the second decade of the twenty-first century will be 'delivered' the benefits of a new official target. The year 2012 is the deadline for crematoria in England and Wales to halve the amount of mercury escaping into the atmosphere from their clients' fillings. It may be the first policy to confound a basic rule of the market. When a waste product is regulated, and thus made harder and more expensive to get rid of legally, the result in the past has always been an increase in fly-tipping. 'Coming to a ditch near you' could pass as the preamble to most European waste directives. One hopes that the amended *Secretary of State's Guidance for Crematoria*, issued under Regulation 37 of the Pollution Prevention and Control (England and Wales) Regulations 2000, will be the first exception.

The poisonous effluence of our rotted teeth is a small example of a larger truth. Waste is as necessary to life as air and water. Just to be born into our own bodies is to create lifelong problems of disposal. Waste, after all, is what life itself becomes when it's over. The invitation to write this book came after I had written for the *Sunday Times Magazine* a pessimistic piece about the looming problems of the national garbage mountain.

Much of what follows is about that same issue, and echoes the horror of everyone from Greenpeace to House of Commons select committees. The cover-notes assert that what I have written is 'often funny'. So it may be, but the laughter is of the nervous kind, a thin lubricant in the jammed mechanisms of disbelief.

Rubbish is more than just discarded waste. It is the hideous inflictions on land and sea of the common agricultural and fisheries policies; the rubbishing of historic town centres by criminally complacent local planning committees; the grid-patterned, rubber-stamped architectural wastelands that people have to live in and look at; the sacrifice of deeply rooted local communities to superficial, tick-box ideologies; the persecution of everything wild and untameable; the penny-pinching, self-harming neglect of the railways and public utilities; the hatred of history; the demeaning of political language with its ludicrous lexicon of holistic stakeholders and sustainable delivery; the smug god-awfulness of television. All these find a place in the story. And yet . . .

Let us not forget our amalgam fillings, or the dust and ash we shall become when the clock ticks its last. To create, to *be*, rubbish is as basic a freedom as speech, or education, or water. Without the freedom to write bad books for the pulping plant, there would be no good ones to be kept in the library. There is no apple without a core; no nourishment without the consequent flush of a cistern; no shelf-life without a sealed and stackable pack; nothing made or grown without its necessary margin of scrap. The economics of waste are often the negative image of what we might expect. Even in basic terms of energy and materials, 'excessive' packaging may save more than it costs. Local authorities with apparently good recycling records may depend for their green credentials on energy from garbage incinerators, or low-grade compost used in landfills. 'Green' initiatives by governments may be a shade so dark that they are indistinguishable from black.

Yes, the fridge mountain was a bit of a laugh; but it wasn't a joke. The same goes for the 'disappearance' of toxic waste after a long-awaited but still unprepared-for change in European landfill regulations. Unless you have shares in Tesco, there is nothing funny about what has happened to the farmed landscape or the high street. There is no humour in a fishless sea. The book is not, or tries not to be, a counsel of despair – that in itself would be a waste of ink and paper. But its optimism is hard won, and based upon the possibly naive assumption that, having allowed problem to become crisis, the government will not allow crisis to become catastrophe.

CHAPTER ONE

History

YOU HAVE TO WONDER ABOUT HISTORY'S SENSE OF SMELL. It was fine in the beginning, when our earliest forebears came down from the trees, gave up being apes and set off across the savannah in search of god. There were not many of us about then. Earth and sea were infinite, like the heavens. We no more had the power to despoil or to desecrate them than we had the ability to change our own natures. There was no need for thought: driven by instinct, we did what came naturally. Our food waste rotted harmlessly where we left it, feeding the soil from which it came. Our dung and urine were dropped like animals', wherever we felt the need. We abandoned our dead to beak and claw and lived like scavengers on our senses. Selfishness and greed were the tools of our survival. Not knowing where our next mouthful would come from, we insured our futures by packing our bellies with everything they would hold. Hunter-gathering was a tough calling. To consume was to live another day, survive another night and pass on our genes. Altruism, like a fatherless child, had a hard start in life.

Years passed in their thousands, and still it didn't matter. If a cave or a patch of land had been fouled, a travelling group would simply move on and find another one. There always *was* another. No need to think about anything but ourselves. No responsibility for, nor any concept of, anything that might be called 'the environment'. Only when we congregated in permanent settlements, and learned to put roofs over our heads, did we have to meet the challenge of living with our own stink. In a way, we were rather good at it. If excrement was a fact of life, then so was the smell of it in the gutters, and who were we to turn up our noses? When food and industrial waste threatened to overwhelm us, we hit on the policy that has served us ever since: chuck it in a hole in the ground. It worked and, in the long-term accumulation of its effects, it has left us both a gift and a curse.

The gift is to archaeologists, whose sifting of pits and middens has told us much of what we know about our ancestors' lives. By their leavings, from horse-bone to silver teaspoon, do we know them. The curse is in our natures. We may have the stink under control but we have not buried the profligacy of our habits. Every hour in the UK we throw away enough garbage to fill the Albert Hall. A gathering tsunami of rubbish – organic and inorganic, active and inert, electronic, aural and visual – pours into our lives and out again, into a world no longer infinite. No more do prophets look to the Bible for signs of The End. We carry Armageddon in our shopping bags. It is not just environmental pressure groups, supposed enemies of modern life, who argue that the planet can no longer afford what we are costing it. In universities and centres of government – even in company boardrooms – there is recognition that we have exceeded our credit limit. We can no longer cope with, or afford, the volume and toxicity of our own wastes. We know we have to change, and yet restraint – the idea of taking less than we can grab – puts us in conflict with the very essence of our genetically driven urge to consume.

Waste has always been a badge of affluence; every possession a piece of junk in waiting. Unsurprisingly, the earliest waste-disposal systems belonged to the people who had the most to throw away. At the Minoan palace of Knossos in Crete, as early as 3000 BC, landfill sites layered with earth were taking the strain of an abundant royal lifestyle. Two-and-a-half thousand years later came the first municipal rubbish dump, a mile outside Athens. Recycling had an early start, too. Uneaten greenstuff was fed to animals, whose own waste fertilized the soil; and Bronze Age Europe well understood the importance of scrap. Britain's first dustmen were the Romans. Addicted to order if not to hygiene, they were as fastidious as they knew how to be. Although they did not follow the ancient Athenians in carting their garbage out of town, they at least buried it in pits. Like modern waste contractors they knew the value of a hole in the ground; and like modern municipal leaders they thought more of public services than they did of high culture. How many city fathers now would love to follow the example of Roman St Albans and dump their rubbish in the theatre? For sewerage, the trailblazer seems to have been the city of Lincoln, which had a proper street-by-street network connected to individual houses. Other towns scent-marked the future, and set the standard for centuries to come, by swilling everything into the street.

If cholera and plague had gods to look after them, then

medieval England was the answer to their prayers. Never has history offered a richer or longer lasting playground for disease. In terms of hygiene, medieval England would stretch well into the nineteenth century, where it would find Charles Dickens waiting for it. Anyone who has felt the need of a lavatory, in a place where none exists, can imagine the state of England's cities. To some extent it was an issue of social class. At the top of the lavatorial hierarchy, thirteenth-century castle-folk in their garderobes voided themselves directly through a flue into the moat (an improvement over the eleventh-century prototype that let everything trickle down the wall). In towns, well-to-do professional and merchant families had scaled-down, middle-class versions falling into streams and rivers. But the common folk shoehorned into their tenements were not so lucky. They had either to queue for shared privies (and you can imagine the state of those) or walk to a public latrine.

Sanitation was seldom a priority for slum landlords, and many provided none at all – not even a bucket. In one typical case in 1421, it was recorded that 'all the tenants threw their ordure and other horrible liquids before their doors, to the great nuisance of holy church and of passers-by'. A century-and-a-half later, little had changed except, perhaps, the sharpening of the stench. In 1579, 85 people living in 57 households in London's Tower Street had only three privies between them. In affairs of the gut, necessity is all. Not every-one with a seething mutton pie inside them could be bothered, or had the time, to find their way to a public latrine. What horses did in the street, so did their human masters.

More caring individuals headed for running water. In London the Walbrook, a natural stream through the middle of the city, was a public sewer into which householders, carters

4

and tradesmen tossed whatever they wanted to get rid of, and over which richer citizens built their privies – every act of nature a gift to the downstream neighbours. Everyone else had to improvise as best they could. Families in top-floor tenements, far from any privy, could hardly be expected to resist the temptation of the window, while others showed varying degrees of inventiveness in fouling anyone's doorstep but their own. In 1314, a woman named Alice Ward was ordered by the mayor to dismantle a wooden pipe connecting her 'privy chamber' at Queenhithe to the gutter in the street below. A few years later, in 1347–8, an Assize of Nuisance dealt with the case of two men who had piped their sewage directly into the cellar of their downstairs neighbour. Not even this was the worst of it. At about the same time, according to Ernest L. Sabine in his classic paper of 1934, 'Latrines and Cesspools of Medieval London', 'certain citizens . . . had constructed divers latrines, in Ebbegate upon gratings, and in Dowgate projecting beyond the pathway, so that in each lane the filth fell upon persons passing through'.

Where there was no running water close by, better-off householders would dig cesspools to act as holding tanks. These brought horrors of their own – some of them immediate and dramatic; others slower and more insidious. Of the immediate kind, no case is more tragic than that of Richard the Raker who, one day in 1326, dropped his breeches and settled down comfortably on his privy. It is hard to imagine a nastier or less dignified way to die. The rotten floorboards collapsed beneath him, plunging him straight into the cesspool, where he drowned in his own waste. Elsewhere, death would come more stealthily in the form of water-borne disease. The worst and most persistent hazard was the leaching of sewage into neighbouring buildings and groundwater, and thence into the wells for drinking. Sabine records a case in 1328–9, when 'Adam Mere and his brother William . . . were summoned before the assize on the complaint of William Sprot that they had a cloaca next his tenement, which was full of filth to over-flowing, so that the dung together with the urine from the cesspool penetrated his wall, entered his house, and collected there, making a great fetor.'

Great fetors throughout the city must have been as hard to escape as the swarms of flies. London's other stream, the Fleet, though wider and deeper than the Walbrook, was in much the same state of degradation, and the Thames itself was the immediate destination of what both streams carried. Where public latrines existed (on London Bridge, for example), they discharged straight into the water, making it less an artery of trade than a tideway of filth carrying the raw sewage of forty thousand people. The better class of citizen did take the trouble to have their cesspools emptied by privy cleaners and taken away by dung-boat, but others – lacking either the means or the will – simply dumped this and every other kind of muck in suburban lanes or along the banks of the river.

Muck there was in plenty. Horses fed dunghills that were frequently big enough to obstruct the highway, and were re-inforced in their endeavours by dung and urine from the city's pigs, cattle and poultry. It might have been possible to pass beyond the reach of putrefaction but it would not have been easy. The stink from discarded butchers' offal, never mind the fishmongers', would turn even stomachs accustomed to rotting meat. It was not just vileness that was the problem. There was the sheer bulk of the city's leavings: straw, sawdust, rushes (used, in vast quantities, to carpet the floors), earth from

cesspools and other diggings, builders' rubbish, dead dogs –
everything the city was unable to eat, sell or recycle.

It was not that nobody in authority noticed or cared. Stung
by complaints from the king, the mayor and city authorities
issued ordinance after ordinance, banning one kind of mis-
behaviour after another and forever seeking new remedies. In
1312, fishmongers were ordered to throw their dirty water into
the Thames and not into the street; in 1366, poulterers were
banned from plucking chickens in the highway. In 1343, under
threat of prison if they failed, the aldermen were commanded
to ensure that the streets were cleared of dung and rubbish. The
job was done by city 'rakers', supervised by 'scavengers' who
carted the stuff away in tumbrils. People who left rubbish in
front of their houses were to be fined – a fact that might
explain a violent incident in 1326, when a pedlar dropped a
couple of eel-skins and an apprentice rushed out of a nearby
shop and struck him dead.

Although there was no real understanding of the vectors of
disease, there was a vague realization that stench, like pain,
was one of nature's ways of steering us away from trouble, and
that bad smells were bad news. Thought and deed, however,
enjoyed only the most tenuous of relationships. In the plague
year of 1349, Edward III wrote to the mayor 'protesting that
filth was being thrown from the houses by day and night, so
that the streets and lanes through which people had to pass
were foul with human faeces, and the air of the city poisoned
to the great danger of men passing, especially in this time of
contagious disease'. The king tried again in 1357, this time
complaining of 'noisome filth' in the city streets and along the
Thames. Fly-tippers now faced heavier fines than ever, and it
was made illegal for anyone to throw rubbish, earth, gravel or

dung into a waterway. What weighed more heavily with many people, however, was the imposition of fees at the official dumps beside the river. Resentment was such that house-holders preferred to risk the law, and carry on dumping illegally, rather than submit to the charges. This drew from the king a further writ banning dumping on Tower Hill, which was answered by yet more dumping in the river, which in turn pro-voked another ban from the city. The suburb of Westminster meanwhile complained of Londoners sending out cartloads of refuse and dung to besmirch its fields and streets.

Rats loved it. The plague returned again in 1361, and again in 1369, 1370, 1382, 1390, 1391 and 1407. It was a vicious circle of the worst kind. Disease made clear the need for better hygiene, yet the epidemics caused such chaos in city govern-ment that, in the years when the need was most acute, the squalor could only get worse. The city authorities did go on trying. First (in 1383) they imposed a levy of two shillings a year on householders with privies over the Walbrook; then (in 1462–3) they banned them altogether. After the plagues of 1390 and 1391, Richard II (a king with much else on his mind) decided that anyone dumping rubbish in the Thames should be liable to a ruinous fine of forty pounds. The one exemption was granted to the butchers. So foul was their rotting offal, and so anxious the king and everyone else to be rid of it, that they were allowed to drop it into the ebb tide. Sabine argued that the never-ending flow of ordinances, levies and prohibitions meant that, within the parameters of its own understanding, medieval England did all it could to keep its head above the filth. You could go further and argue that its public policies were closer to the known 'best practice' of the age than they have been so far in the late twentieth and early twenty-first centuries. What you could not convincingly argue is that they worked. Kings came and went. So did mayors. Years rolled on into decades, and decades into centuries, yet still the streets swilled in waste and the population remained a soft target for any disease that liked a bit of dirt.

The mysterious 'English sweating sickness', known also as *Sudor anglicus*, struck first in 1485, then again in 1508, 1517, 1528 and1551. Attributed variously to lice and ticks, or simply to 'filth', and possibly caused by a virus, it was as unforgiving as the plague itself. Death was preceded by headaches, high fever, muscle pain, profuse sweating, skin rash and laboured breathing. The court of Henry VIII was particularly hard hit, and the king himself took especial pains to avoid contact with it. In 1528, wrote the French ambassador, 'One of the *filles de chambre* of Mademoiselle Boleyn was attacked on Tuesday by the sweating sickness. The king left in great haste and went a dozen miles off . . . This disease is the easiest in the world to die of. You have a slight pain in the head and at the heart; all at once you begin to sweat. There is no need for a physician: for if you uncover yourself the least in the world, or cover yourself a little too much, you are taken off without languishing. It is true that if you merely put your hand out of bed during the first 24 hours . . . you become stiff as a poker.' No sooner had *Sudor anglicus* paid its last recorded visit in 1551 than bubonic plague reappeared to assert its place as reaper-in-chief. Rat-borne fleas cost the lives of 17,500 Londoners in 1563; 23,000 in 1593; 30,000 in 1603; 40,000 in 1625 and 80,000 in 1665.

While London during the eighteenth century grew exponentially in size and elegance, the old problem was never far from sight. Here is Lord Tyrconnel in 1741, quoted by Roy Porter in *London, a Social History*, denouncing the 'neglect of cleanliness of which, perhaps, no part of the world affords more proof than the streets of London, a city famous for wealth, commerce and plenty, and for every other kind of civility and politeness; but which abounds with such heaps of filth, as a savage would look on with amazement'. Others complained of ordure lying in the streets, and of roads lost beneath stagnant lakes of liquid mud. Elsewhere it was much the same. Other towns, too, were wallowing in filth. The 1750s brought complaints of dung heaps in the streets of Southampton – this despite Messrs Warwick and Minshaw paying the mayor annually ten guineas plus a brace of capons in return for the right to collect the city's waste. If muck has often meant brass, it has just as often meant corrupt deals and opportunism, as well as a thin trickle of creative genius. In 1776, America discovered the value of recycling when New Yorkers melted down a statue of George III to make bullets.

Accounts of nineteenth-century London were very little different from those of half a millennium earlier. Porter quotes the surgeon John Simon, appointed Medical Officer of Health in 1848: 'Let the educated man devote an hour to visiting some very poor area in the metropolis. Let him fancy what it would be to himself to live there, in that beastly degradation of stink, fed with such bread, drinking such water . . . Let him talk to the inmates, let him hear what is thought of the bone-boiler next door, or the slaughter-house behind; what of the sewer-grating before the door; what of the Irish basketmaker upstairs – twelve in a room; what of the artisan's dead body, stretched on his widow's one bed, beside her living children.'

It was cholera now that made merry, fattening itself in

polluted wells – wells whose bright water, made sparkly by
ammonia and other organic contaminants, suckered its victims
with eye appeal. From out of squalor and suffering, however,
came the first glimmerings of lasting hope. The rapid development of rational science and medicine, and the technological
red heat of the industrial revolution, brought forth a generation
of men – self-important, perhaps, but possessed of a furious
energy – who recognized the essential need for social as well
as economic and industrial change. The motivators were
philanthropy, horror, and the recognition that better health was
an essential precursor of improved national prosperity.

The horror was genuine. In the summer of 1842 the
secretary of the Poor Law Commission, Edwin Chadwick,
presented to the House of Lords his *Report on the Sanitary
Conditions of the Labouring Population of Great Britain*. It
was, in effect, a nationwide gazetteer of degradation,
putrescence and disease. You could look almost anywhere and
find the same. This account, received by Chadwick from Mr
Bland, medical officer of Macclesfield, is typical:

In a part of the town called the Orchard, Watercoates, there are
34 houses without back doors, or other complete means of
ventilation; the houses are chiefly small, damp, and dark; they
are rendered worse with respect to dampness perhaps than they
would be from the habit of the people closing their windows to
keep them warm. To these houses are three privies uncovered;
here little pools of water, with all kinds of offal, dead animals
and vegetable matter are heaped together, a most foul and
putrid mass, disgusting to the sight, and offensive to the smell;
the fumes of contagion spreads periodically itself in the
neighbourhood, and produces different types of fever and

disorder of the stomach and bowels. The people inhabiting these abodes are pale and unhealthy, and in one house in particular are pale, bloated, and rickety.

Nineteenth-century Macclesfield or fourteenth-century London? Who could tell the difference?

A couple of pages further on, we find Mr Pearson, medical officer of Wigan:

Many of the streets are unpaved and almost covered with stagnant water, which lodges in numerous large holes which exist upon their surface, and into which the inhabitants throw all kinds of rejected animal and vegetable matters, which then undergo decay and emit the most poisonous exhalations. These matters are often allowed, from the filthy habits of the inhabitants of these districts, many of whom, especially the poor Irish, are utterly regardless both of personal and domestic cleanliness, to accumulate to an immense extent, and thus become prolific sources of malaria, rendering the atmosphere an active poison . . . It may be also mentioned that in many of these streets there are no privies, or, if there are, they are in so filthy a condition as to be absolutely useless; the absence of these must, necessarily, increase the quantity of filth, and thus materially add to the extent of the nuisance.

Here is Mr Rowland of Carlisle: 'on the south side at the foot of Botchergate, there is a gutter, perhaps a mile long, which conducts the filth of that quarter through the fields into the river Petteril. The stench in summer is very great.'

Mr Aaron Little, of Chippenham, on the rural parish of Colerne: 'The filth, the dilapidated buildings, the squalid appearance of the majority of the lower orders, have a sickening effect upon the stranger who first visits this place. During three years' attendance upon the poor of this district, I have never known the small-pox, scarlatina, or the typhus fever to be absent . . . There is also a great want of drains.'

Mr Parker, of Windsor: 'From the gas-works at the end of George-street a double line of open, deep, black, and stagnant ditches extends to Clewer-lane. From these ditches an intolerable stench is perpetually rising, and produces fever of a severe character.'

Dr Edward Knight of Stafford: 'There is not any provision made for refuse dirt, which, as the least trouble, is thrown down in front of the houses, and there left to putrefy.'

Mr William Rayner of Stockport: 'The street . . . is seven yards wide, in the centre of which is the common gutter, or more properly sink, into which all sorts of refuse is thrown; it is a foot in depth.'

Mr Robert Atkinson, of Gateshead: 'It is impossible to give a proper representation of the wretched state of many of the inhabitants of the indigent class, situated in the confined streets called Pipewellgate and Killgate, which are kept in a most filthy state, and to a stranger would appear inimical to the existence of human beings.'

The Reverend Dr Gilly, canon of Durham, on a peasant's hovel: 'It is not only cold and wet, but contains the aggregate filth of years, from the time of its first being used. The refuse and dropping of meals, decayed animal and vegetable matter of all kinds, which has been cast upon it from the mouth and stomach, these all mix together and exude from it.'

And so on, in town and village throughout the country, each local official struggling to convey the full horror of what he has seen, many of them apparently believing their situation to be so bad that it must be unique. The true horror, recognized by Chadwick and other reformers of the time, was that the squalor was ubiquitous. Even badgers cleaned their dens. We were worse than animals, preferring cholera and typhus to the chore of removing our rubbish, as ready to waste years of our own lives as to throw down the carcass of a rabbit. Between 1848 and 1854, the death toll from cholera alone was a quarter of a million, and 15 per cent of children did not survive their infancy. It was not just more regulation that was needed – there had been no shortage of that – but a complete re-ordering of local and national priorities. Though many of his own ideas were impracticable (the basis of his proposed economic miracle was the export of metropolitan sewage for use as agricultural manure), Chadwick himself was prominent in the clamour for reform. His *Report* caused deep shock, as he intended it should, and he took care to send copies to opinion-formers such as John Stuart Mill and Charles Dickens, whose last completed novel, *Our Mutual Friend*, is rooted in the black economy of the dust-yard.

One small but vital step forward came with Britain's first Public Health Act in 1848, which established a General Board of Health and handed to corporate boroughs the responsibility for drainage, water supply and 'removal of nuisances' (London got its own City Sewers Act). In Dickensian London, waste disposal was already an engine of profit. In his massive four-volume survey of 1861, *London Labour and the London Poor*, the maverick journalist and co-founder of *Punch* Henry Mayhew provides an obsessively detailed description of dust-men's lives, including everything from their wages to their drinking and sexual habits (dust*men* tended to live with dust*women*, with whom they would have dust*children* and create entire dust dynasties). Muck, as ever, was brass. The contracts for refuse collection in London's 176 parishes were shared between 80 or 90 contractors, who also had responsibility for cleaning the streets – contracts which, in Mayhew's estimation, were worth in total between £30,000 and £40,000 a year. Though modern waste disposal contractors would find it extraordinary *not* to be paid for their work, in the mid nineteenth century the idea was something of a novelty. Initially it was the contractors who paid the parishes for the right to cart away their dust (chiefly composed of ash) and to profit from its sale, either as fertilizer or for mixing with clay in brickmaking. As demand for dust declined, so the balance changed. The parish of Shadwell, for example, having once received £450 a year from its contractor, now had to pay him £240 to take the stuff away.

Mayhew's description of the collection round is a model of documentary precision. Two men – a 'filler' and a 'carrier' – tour the streets with 'a heavily-built high box cart, which is mostly coated with a thick crust of filth, and drawn by a clumsy-looking horse'.

These men used, before the passing of the late Street Act, to ring a dull-sounding bell so as to give notice to housekeepers of their approach, but now they merely cry, in a hoarse un-musical voice, 'Dust oy-eh!' The men's equipment consists of a short ladder, plus two shovels and baskets. These baskets one of the men fills from the dust-bin, and then helps them alter-nately, as fast as they are filled, upon the shoulder of the other man, who carries them one by one to the cart, which is placed immediately alongside the pavement in front of the house where they are at work. The carrier mounts up the side of the cart by means of the ladder, discharges into it the contents of the basket on his shoulder, and then returns below for the other basket which his mate has filled for him in the interim. This process is pursued till all is cleared away, and repeated at different houses till the cart is fully loaded; then the men make the best of their way to the dust-yard, where they shoot the contents of the cart on to the heap, and again proceed on their regular rounds.

Mayhew calculated that most two-man teams would bring back five cartloads a day. In the yard itself, the dust was sifted and sorted in a treasure hunt of heroic energy. As well as the *fillers* and *carriers* who brought in the carts, four categories of worker were employed by the contractor: a *yard foreman* or superintendent; *loaders* of outgoing carts; *carriers* of cinders or bricks to their respective heaps; and a *foreman* or *forewoman* of the heap, also known as *hill-man* or *hill-woman*. This last was a powerful figure who employed yet more labourers of his or her own. By Mayhew's account these were arranged into three more categories:

1. *Sifters*, who are generally women, and mostly the wives and concubines of the dustmen, but sometimes the wives of badly-paid labourers.
2. *Fillers-in*, or shovellers of dust into the sieves of the sifters (one man being allowed to every two or three women).
3. *Carriers off* of bones, rags, metal, and other perquisites to the various heaps; these are mostly children of the dustmen.

Put all these together and you construct a scene worthy of

16

Pieter Brueghel. A medium-sized yard would need perhaps twelve collectors, three fillers-in, six sifters and one foreman or -woman; a large one might need a workforce of 150. Mayhew's own account is unimprovable.

Near the centre of the yard rises the highest heap, composed of what is called 'soil', or finer portion of the dust used for manure. Around this heap are numerous lesser heaps, consisting of the mixed dust and rubbish carted in and shot down previous to sifting. Among these heaps are many women and old men with sieves made of iron, all busily engaged in separating the 'brieze' [coarser lumps] from the 'soil'. There is likewise another large heap in some other part of the yard, composed of the cinders or 'brieze' waiting to be shipped off to the brickfields [where it might fetch perhaps three shillings a ton]. The whole yard seems alive, some sifting and others shovelling the sifted soil on to the heap, while every now and then the dust-carts return to discharge their loads, and proceed again on their rounds for a fresh supply. Cocks and hens keep up a continual scratching and cackling among the heaps, and numerous pigs seem to find great delight in rooting incessantly about the garbage and offal . . .

In a dust-yard lately visited the sifters formed a curious sight; they were almost up to their middle in dust, ranged in a semi-circle in front of that part of the heap which was being 'worked'; each had before her a small mound of soil which had fallen through her sieve and formed a sort of embankment, behind which she stood. The appearance of the entire group at their work was most peculiar. Their coarse dirty cotton gowns were tucked up behind them, their arms were bared above their elbows, their black bonnets crushed and battered like those of fish-women; over their gowns they wore a strong leathern apron, extending from their necks to the extremities of their petticoats, while over this, again, was another leathern apron, shorter, thickly padded, and fastened by a stout string or strap round the waist. In the process of their work they pushed the sieve from them and drew it back again with apparent violence, striking it against the outer leathern apron with such force that it produced each time a hollow sound, like a blow on the tenor drum. All the women present were middle-aged, with the exception of one who was very old – 68 years of age she told me – and had been at the business from a girl. She was the daughter of a dustman, the wife, or woman of a dustman, and the mother of several young dustmen – sons and grandsons – all at work in the dust-yards at the east end of the metropolis.

From out of the grey mountain, the sieves would produce all kinds of bits and pieces that had a particular value of their own. Broken bricks, oyster shells and rubble could be sold for laying as foundations under concrete. Rags and bones went for paper-making and glue. Tin and other metals went to make fastenings or 'clamps' for trunks. Boots and shoes were sold to makers of Prussian blue, who had a use for them in the manufacturing process. Money and jewellery, as Mayhew put it, were 'kept, or sold to Jews'.

Even without such 'perquisites', by the standards of the day dust-yard workers were not badly paid. A single man might expect to make fifteen shillings a week, and a 'married' man helped by his family could expect, on average, £1 or more – this at a time when a seamstress would be lucky to make more than sixpence a day, and an agricultural labourer eight shillings a week. This is not to say that the contractors were generous. While they were always cagey about their profits ('they seem to feel that their gains are dishonestly large, and hence resort to every means to prevent them being made public'), they kept tight control over what went into their employees' pockets – even to the extent of deducting from carters' wages the 'perquisites', offered usually in the form of cash or beer, that the men received from grateful householders. This so depressed the men's incomes that they took to demanding the 'perquisites' as a right, making their point by scattering dust, cinders and other rubbish outside the houses of non-payers.

In the kitchens and parlours of better-run homes, waste avoidance was a highly developed art. As Judith Flanders records in her meticulous portrayal of nineteenth-century domestic life, *The Victorian House*, nothing was thrown away that was not beyond all hope of further use. In the kitchen, fish-heads, plate-scrapings and vegetable water went into soups and gravies, and stale bread into puddings. Anything left would be recycled as pigswill. Soiled paper went on to the fire, while clean was torn up either to serve in the lavatory or to be twisted into 'spills' for lighting candles or fires (a habit that persisted in some homes well into the second half of the twentieth century). Worn-out sheets became bandages. Rag-and-bone men took other textiles and bones, and the back door received a steady flow of dealers ready to buy paper, metal and anything else for which human ingenuity could devise a future. Only the careless, the drunk or the profligate would leave very much for the sifters to find at the yards.

'Night soil' was emptied from the cesspools by the same men who collected the dust and cleaned the streets, though by a somewhat different arrangement. For this there were no parish contracts, only private agreements between landlords and the contractors. Anthony S. Wohl, in *Endangered Lives*, suggests it was the very stench of cesspool-cleaning that deterred local authorities from accepting responsibility for it. For extra pay, 'nightmen' – or 'shit-sharks' as they were more popularly known – would go out after dark to perform the noblest of their deeds (luckily, one of the benefits of their calling seems to have been immunity to smells). Not every tenant or landlord, however, could be relied upon to spare the expense. Neither was it always the case that there was a contractor available. Wohl reports, for example, that the entire population of Ipswich – 45,000 people – shared the services of just four cesspool cleaners. Cesspools frequently overflowed or leaked, and the rivers – now also bearing the assaults of increasingly heavy industrial effluent – remained as polluted as ever. Poor drainage meant the consequences of a downpour could be far worse than a drenched hat and coat. Again it is Mayhew who paints the picture:

Until towards the latter end of the last century . . . the streets even of the better order were often flooded during heavy and continuous rains, owing to the sewers and drains having been choked, so that the sewage forced its way through the gratings into the streets and yards, flooding all the underground apartments and often the ground floors of the houses, as well as the public thoroughfares with filth.

It is not many months since the neighbourhood of so modern a locality as Waterloo-bridge was flooded in this manner, and boats were used in the Belvidere and York-roads. On the 1st of August, 1846, after a tremendous storm of thunder, hail, and rain, miles of the capital were literally under water; hundreds of publicans' beer cellars contained far more water than beer, and the damage done was enormous. These facts show that though much has been accomplished towards the efficient sewerage of the metropolis, much remains to be accomplished still.

Neither was it just London's problem. As late as the 1880s one can find descriptions of Cambridge as 'an undrained, river-polluted, cesspool city'. Even at Windsor, the castle sewers would overflow and drench the lawns in excrement rendered no less offensive by its courtly origins. Towns everywhere shimmered in a haze of blended stinks that added rotting vegetable matter, dead cats, animal offal and blood to the gut-wrenching effects of human and animal excrement. According to L. C. Parkes, quoted by Wohl, the poor 'were in the habit of depositing their excreta in a newspaper, folding it up, and throwing it . . . out of the back window'. If the smells were medieval, so too was the prevailing view among medical men that it was from this foul 'miasma' that disease would spontaneously arise.

It was to rid London of the supposedly infectious cloud, rather than intentionally to purify the cholera-infected water supply, that Sir Joseph Bazalgette designed his justly famous London sewerage system. This was begun after the 'Great Stink' had forced the House of Commons to adjourn, retching and spluttering, in June 1858, and was finished in 1875 when it was opened at a grand ceremony attended by the Prince of Wales and the Archbishop of Canterbury. Bazalgette's aim was to get rid of open sewers and to prevent sewage flowing directly into the river, at least within central London (though by the end of the century, according to Wohl, the two outfalls at Barking and Crossness were releasing a torrent of 150 million gallons a day – one sixth of the river's total volume of water). The 'miasma' duly relaxed its grip, but the real benefit was to the city's water supply. London's last serious cholera epidemic was in 1866. Nationwide, however, progress was not so fast. Wohl tells us, for example, that in 1911 two thirds of Manchester's working-class households were still using bucket lavatories, ash-boxes or privy middens. In seeking to explain the high rates of typhoid, scarlet fever and diarrhoea, the local medical officer of health had only to follow his nose.

Other improvements were double-edged. In 1874 Britain got its first prototype waste incinerator, the so-called 'destructor', at Nottingham. Another 250 would be built during the next thirty years – each one a mini volcano deluging its neighbourhood with a sooty lava of ash, dust and charred paper. (Thus began the long and bitter opposition to incinerators that has never ceased.) Victorian rakers would be replaced in the next century by hollow-faced men picking recyclable material from the conveyor belts that fed the furnaces. In 1875 came a further and stronger Public Health Act. Local authorities now had a *duty*, not merely the right, to arrange for the regular collection of household waste, and in 1907 this was extended to include trade waste. In 1898 the business of waste disposal acquired professional gravitas when its senior practitioners formed the Association of Cleansing Superintendents – an organization that would later metamorphose into the Chartered Institution of Wastes Management. It was a worthy body whose good intentions could occasionally lure it into over-optimism. In 1907 it predicted that the greatest advance 'in the near future' would be a change of emphasis from destroying refuse to salvaging it. We are still waiting.

There was a little flicker of hope with the birth, in 1921, of the Association of London Waste Paper Merchants (now transmuted into the British Recovered Paper Association), with its self-interested but nonetheless worthy ambition to recycle more paper. Despite Manchester's allegiance to squalor, more and more households were enjoying the benefits of modern sewerage, flush lavatories and the historic absence of stink. In Burnley, for example, the number of WCs increased from 586 in 1874 to 20,691 by 1900. Robust old slang terms – *jakes*, *bog-house* – would give way gradually to genteel euphemisms: *convenience*, *toilet*. Just over the temporal horizon lurked lavatory cleaners ('kills all known germs!'), air and water fresheners, toilet-roll cosies and an excretal coyness that all but denied possession of kidneys, bladder and bowels altogether. Dustbins made their appearance before the First World War, and people gradually grew accustomed to using them. Horse-drawn dustcarts were being overtaken by motor-driven ones, and many urban streets for the first time flowed with pedestrians and traffic instead of sewage and the effluvia of rotted household waste. An historic problem had finally gone away, if only in the sense of being swept under the national carpet. Out of sight might have meant out of mind, but it did not mean out of existence.

The rubbish still had to go somewhere, which, to the cleansing superintendents of the early twentieth century, meant exactly what it had meant to the Romans – holes in the ground, or 'landfill' sites. These were hard on the eye, even harder on the nose, and carried the age-old threat of polluted groundwater. In 1930 an alarmed Ministry of Health protested that 'the system of dumping crude refuse without taking adequate precautions should not be allowed to continue' – this at a time, in the pre-plastic, pre-chemical, pre-electronic age, when garbage was a much more benign and less volatile commodity than it is seventy-five years later. Most houses were still warmed by open fires, which consumed the bulk of the paper, while the grimy endeavours of the rag-and-bone men kept down the volumes of metal, cloth and glass. Typically in the 1930s the average dustbin would justify its name by containing mostly dust or ash. A. E. Higgins, in *The Analysis of Domestic Waste*, reckoned it at 56.9 per cent. Paper, at just 14.3 per cent, was the next largest single category, followed by vegetable waste at 13.7 per cent, metals at 4 per cent, glass at 3.4 and textiles at 1.9. Compared with the contents of the average modern wheelie bin, this all looks fairly harmless. Even so, there was plenty for hungry bacteria to get busy on and to convert, insidiously over the years, into a dangerous build-up of methane gas. Even now, old landfill sites still remain as potent a threat as landmines or unexploded bombs. As recently as 1986 and 1989, houses at Loscoe in Derbyshire and Kenilworth in Warwickshire were devastated by exploding landfill gas. Even without combustion, methane is an actively dangerous greenhouse gas – one of the worst contributors to global warming, with twenty-one times the destructive power of carbon dioxide.

Even if this had been understood at the time, it would have vanished into insignificance when set against the horrific human and economic waste of two world wars. Though reclamation was an important part of the British war effort, refining the waste-disposal system did not figure highly in the national consciousness. By the end of both wars, the country was scabbed with huge, uncontrollable tips, some of them up to a mile in length. If there was, for a while, a legacy of hope, it lay in the habit of prudence ingrained by rationing and the make-do-and-mend of wartime austerity. In 1955 the county engineer of Dumfries County Council, J. C. Wylie, wrote an almost heartbreakingly optimistic book, *Fertility from Town Wastes*, detailing all the good uses to which garbage might be put. At a time when *Waste Not, Want Not* was not so much a popular motif as an entire philosophy of life, engraved as deeply on the nation's post-war psyche as it was on novelty salt cellars, it looked like a parable of enlightenment. Nothing would go to waste. Even dust – still the principal ingredient of the household bin – might be sold for brickmaking or fertilizer.

'Refuse dust is composed largely of carbonaceous matter, a large proportion of which consists of fine particles of partially burned coal. It is known that coal is a valuable source of plant nutrients.' Tin cans were baled and processed through scrap merchants, 'upon which the steel industry of the country depends for supplies'. Bones, too, still had a high commercial value provided they were clean (otherwise they 'quickly cause offence when stored'), when they could be 'pulverised and incorporated with other organic wastes to produce composts. Pulverised bones are not discernible in mature composts while their presence adds small but useful amounts of organic phosphorus.'

Such economies were the physical manifestations of a public mood – like an endless recycling of traditional family values. The Board of Trade, forerunner of the Department of Trade and Industry, even set targets for the collection of 'vegetable matter' – one ton per thousand head of population per month, a target comfortably exceeded in, for example, the London borough of East Ham, where specially designed vehicles called three times a week to collect the peelings and convert them into 'a substantial source of income' for the local council. In due course the food value extracted from them would reappear on the family's table in the form of pork and bacon from the pigs that had been fed on it. Some local councils even went a logical step further and, instead of selling to farmers, set up their own piggeries. In 1950–1, Wylie reported, the municipal piggery in Aberdeen made a profit of £3,000 (£67,800 in today's terms).

To read Wylie is to imagine a world made wise by suffering, now pursuing a waste-free utopia in which prudence would become the pathway to plenty. *Waste Not, Want Not.* The tragedy of ideals is that they inhabit the minds of idealists. Wylie's ideas fitted the world he knew, but the world he knew was yesterday's. Already the future was closing its grip. Paper, for example, was starting to become a problem – a problem that would worsen as the 1956 Clean Air Act damped down the fires and hastened the switch to central heating. The production of plastics from petroleum-derived chemicals had started in the 1930s but now began to rise sharply, the perfect metaphor for the belated arrival, at full strength, of the twentieth century. People had eaten their fill of prudence and austerity, and it had left them thirsting for change. They had thrown away their ration books; raised their hems, their horizons and their expectations. They wanted to *consume*. In the world of the clam-pack, Wylie's dust-based economy harked back to the age of the woodcut and the homeward-wending ploughman. Except ... the new challenge brought forth only the same old answer. The explosion in consumption, and the Vesuvian eruption of rubbish it unleashed, was to be swallowed by holes in the ground.

Waste policy, insofar as there was one, was driven by the politics of disgust. Landfill fouled the groundwater and spiked itself with methane. If anyone argued for the precautionary principle, their voices went unheard. Not until disaster had already struck did anyone seek to avert it. In 1960, when a working party set up by the Duke of Edinburgh flared its nostrils at the deteriorating state of the environment, we got the Royal Commission on Environmental Pollution. In 1972, after some drums of cyanide waste had been dumped in an old brick kiln near Nuneaton, we got the Deposit of Hazardous Waste Act. We got bottle banks. We got European waste directives. We got dumped on.

We got private contractors in place of the corporation dust-cart. We got golf courses, housing estates and shopping centres 'landscaped' with black-market garbage. We got fields, ditches, verges and lay-bys swarming with fly-tippers; bridle-ways blocked with the rusting cadavers of Lady Thatcher's 'great car economy'. We got town councils fiddling the re-cycling figures; the Department of Trade and Industry deciding that electricity generated by filthy municipal waste incinerators should count as 'renewable energy'; carcinogenic ash spread on children's playgrounds, footpaths and allotments, and more of the same built into people's homes. Every available hole in the ground overflows with yesterday's rubbish and no room is left for tomorrow's.

Like our simian-browed ancestors of the African savannah, we waste without constraint. Pigs haven't seen swill since 2001 (when it was blamed for an outbreak of foot-and-mouth disease). We live in the chew-and-spit age of bubble-gum and the text message, when even language is dispensable. Junk is not so much a by-product of modern life as the foundation of it. We have a daily diet of junk mail, junk food, junk emails, and junk celebrities whose fleeting hours of fame are created by junk television. In a throwaway culture, nothing is designed for the future. It is fashion, not utility, that governs the lifespan of what we buy. Good, workable mobile telephones become junk as soon as manufacturers launch their new ranges. Clothes become junk twice a year, in spring and autumn, as fashion changes its mantle. Music becomes junk the moment we've heard it. Waste is endemic in our systems of government, law and education. For everything we buy, there is something to throw away. Empty buildings are waste. A crumbling rail system is waste; leaking water mains are waste; traffic jams are waste; subsidized agricultural over-production is waste. Early deaths from bad diets or bad habits are waste. Copycat programming on multiple television channels is waste.

But waste is sexy. Fashion depends on it. Pleasure, almost by definition, thrives on inessentials, not on organic bread and filtered rainwater. Conservationists, with their environmental audits and sustainability fetish, are the new puritans. They are enemies of choice, zealots in big knickers who would have us recycle our own ordure before they would let us eat a Big Mac. Their determination to prevent us doing what we want increases in direct proportion to the strength of our need to do it. It is as unreal as alchemy, as if they think there's a recipe for turning base humanity into self-denying nuggets of moral gold. To deny us our appetites is both an offence against nature and an infringement of our liberty. We do not want to go every-where by bio-fuelled bus, or wear clothes made out of recycled hamburger packs, or live on roots and pulses. We want the right to choose, and, where the choice is between sex and sustainability, no one should expect an overwhelming vote for virtue.

CHAPTER TWO
Water

TAKE AWAY ALL BACKGROUND NOISE. SILENCE THE WIND. The one sound you're left with is the sound of running water. It burbles in sewers and water mains beneath our feet; hisses in the tanks above our heads; hurtles down mountainsides in gunmetal threads, sharp as cheesewire, that have cut the landscape since the last ice age. It fills the rivers and the reservoirs, moistens a land of poetically famous verdure, and finally spills out into a grateful sea. Even then its journey isn't done. In a process older than life, it begins again the perpetual cycle of transpiration that will convert it from liquid into vapour, and from cloud back again into the rain that never seems to stop. Along the river valleys, large parts of Britain live in permanent fear of flood.

All this is true. It is also a grand illusion. Parts of England have less water per head than Ethiopia or the Sudan. England and Wales on average have less than any country in Europe except Belgium and Cyprus – less even than the dusty hinterlands of Spain and Portugal. After centuries of clutching our stomachs, holding our noses and burying the prematurely dead, we have finally learned to treat it with a little more respect. The Prince of Wales may have cause to complain about the pollution of groundwater by agro-chemicals, but it is nothing to set against the horrors witnessed by the Houses of Plantagenet, Tudor, Stuart and Hanover. Raw sewage in river or sea, though far from unknown, is now the exception rather than the rule.

Without *treated* sewage, however, much of the inland water-way system would simply dry up. The summer flows of many rivers in southern England, for example, depend entirely on the outflows from sewage treatment works. The volumes are so vast they are beyond imagining. At an individual level, you can just about grasp it – the average water consumption in the South-East is about 383 litres (84 gallons) per household per day, with the same amount being sent back via the drains in the form of grey water and sewage. Across the UK, this grosses up (and, for much of it, 'gross' is the right word) to 11 billion litres (2.42 billion gallons) – enough to float 189 aircraft carriers the size of *Ark Royal*. Every day we get through 13m rolls of lavatory paper. Unrolled and laid end to end, this would stretch for 400,000 kilometres, or 250,000 miles, enough for ten complete loops of the earth. As a nation we are almost manically obsessed with our bowels, forever adjusting our personal outflows up or down. In 2002, in pharmacy and grocery stores alone, we spent £36.2m on over-the-counter remedies for diarrhoea, and £40.8m on laxatives. GPs write 10m prescriptions for laxatives every year. Our colons and urethras are windows on our inner being: the NHS can't say how many samples of faeces and urine it receives each year, but the figure must run deep into the millions. Contrarily, how-ever, where once Merrie England wallowed in its own ordure, we now have such a mordant distaste for our own by-products that it borders on the neurotic. Building regulations routinely insist on bathroom extractor fans to relieve us of our personal miasma. From corner shop to out-of-town superstore, endless shelf space is devoted to a multiplicity of products designed to eliminate 'germs' and smells: bleaches, disinfectants, special cleansers, scented water and air fresheners.

Product labels are slaves to the ideal of 'natural freshness', as if each ceramic pedestal were the source of a crystal mountain stream. In terms of hydrology at least, the image is surprisingly apt. Tiny, personal trickles conjoin and swell into bigger and bigger tributaries until they become, if not a raging torrent, then at least a deep and powerful surge. The basic technology was understood even before Sir Joseph Bazalgette designed his scheme for London in the late 1850s (though it was another Briton – the Scotsman James Newlands, appointed Liverpool borough engineer in 1847 – who was credited with the world's first purpose-built sewerage system). The UK now has 347,000 kilometres (215,600 miles) of sewers, connecting to some 9,000 treatment works that process the sewage to varying standards of purity, then pump the treated liquid into stream, river, estuary or sea, and spread the solids on farmers' fields to plump up our Brussels sprouts and lettuce. The ghost of Edwin Chadwick might well ask: who's laughing now?

One statistic would come as no surprise, even to a Roman or a fourteenth-century Assize of Nuisance: on average, 98 per cent of what the sewers convey is water. From sink, lavatory and washing machine it streams out of our houses. By way of roof, road and gutter it pours down from the sky. In a myriad different colours it foams and seethes from workshops, factories and laundromats. Schools, hotels, restaurants, hospitals, offices, prisons, the Houses of Parliament, Buckingham Palace, football grounds, cinemas, motorway service stations – all are ceaselessly engaged in a national tattoo of clank and whoosh. However, water may be 98 per cent of the traffic but it's a much smaller fraction of the problem. It's the other 2 per cent that we have to fear.

Even in passing water through our bodies, we pollute it in more ways than we may realize. Bacteria, yes, we know about those. That's why we keep soap in the lavatory. It's why the House of Commons Public Accounts Committee, during its examination of the Food Standards Agency in 2002, was shocked when it heard that half the catering staff in England and Wales do not wash their hands before preparing food, and that a third don't wash after using the lavatory. *But eating vegetables, taking our medicine, washing our hair, cleaning our homes, going to the dentist?* What possible harm could any of those do? The answers are intriguing. Vegetables absorb metals – cadmium, for example – from the soil. The amounts are tiny, but each of us passes on our mite and, by the time the flushes are pooled at the treatment works, it all adds up. Pharmaceutical residues go the same way: antibiotics, analgesics, anti-depressants, sleeping pills, anti-coagulants, steroids, hormone-disrupters – whatever our bodies have not absorbed, we expel to become yet another ingredient in the national soup. Hygiene in its aftermath is as filthy as health. Anti-dandruff shampoos contain zinc; so do many cosmetics and hair-care products. Hair-lice preparations contain what chemists call 'complex organic molecules'; so do detergents, household cleaners, garden pesticides and weed killers. When we rinse our mouths after a filling at the dentist, we spit out

WATER

fragments of mercury – one of the most toxic substances covered by the European Dangerous Substances Directive, which is supposed to staunch the flow into the environment of any material likely to harm it. Even new domestic plumbing for a while adds copper to the flow. All this stuff, and more, we rinse and flush without restraint into the public sewers.

The water companies – whose responsibility it becomes as soon as it crosses the boundary of our homes, and who must clean up the water before they release it back into the environment – have no more control over what we send them than they have over the phases of the moon. Nor do they have any control over how many of us plug into their systems – we all have a free statutory right to connect to the nearest public sewer, and the industry has no reciprocal right to be involved in the planning process. If the Office of the Deputy Prime Minister decides to drop into the south-east of England new developments equivalent to the size of Leeds, then the water companies will simply have to cope, and 700,000 extra sets of bladders and bowels will have to be catered for.

There is another, more literal sense, too, in which the problem is not easily soluble. Into the swill goes a jostling flotilla of inorganic solids looking for pipes to block: sanitary towels, tampons and their plastic applicators, panty liners, disposable nappies, baby wipes, incontinence pads, tights, condoms and femidoms, colostomy bags, bandages, plasters, cotton buds, toothbrushes, cleaning rags, razor blades, even hypodermic syringes and needles – anything at all that we hope will navigate the U-bend. Some of us are not content even with that. What to do with the old, blackened sump oil from the car? Up with the nearest manhole cover, and down the hatch. Down, too, go unused paint, dead pets, plastic bags,

33

supermarket trolleys, bicycles, old motorbikes. The only wonder is that the system – much of which is more than a century old – doesn't clog up more often than it does. When a sewer becomes blocked or flooded, the result is far beyond the odourless zone of everyday nightmare. It is just as Henry Mayhew described it in the mid nineteenth century. The sewage under pressure looks for the nearest exit, and up it comes, foaming and alive, through manholes and lavatories, a foul brown surge that degrades everything it touches. Houses, cellars, streets and gardens flood with it. When the tide abates, it leaves in its stinking dun drapery the cruellest exposure of the lie that underpins that weasel word 'disposable'. To those afflicted – and each year it happens to between 5,000 and 7,000 properties in England and Wales – it is the crudest and most unforgettable crushing of human dignity.

Tarmac is another powerful enemy. In open country, rain soaks down through the soil to feed the aquifers and perpetuate the natural cycle. But this cannot happen where there are buildings. Roads, houses, shopping centres, industrial estates, car parks – all combine to form a waterproof barrier between sky and earth, leaving the rain to find other ways to escape. By design, much of it flows into the sewers. Sometimes it has a separate surface-water system all of its own, and travels via pipe and conduit direct to a watercourse and thence to the sea. Sometimes it joins the sewage water on its way to the treatment works. Sometimes, hideously, it does both. In many old 'combined' systems, separate surface-water and foul-water sewers run side by side in the same trench. At various points, as an insurance against overloading, they interconnect so that, if either one overflows, it escapes into the other. The result, when the foul water overflows, is that raw sewage joins the surface water and is borne untreated into river or sea. This is particularly a risk during heavy storms when the foul sewers, too, fill up with rainwater – a risk that will only get worse as global warming drives the climate to ever greater extremes.

One heavily touted alternative, popular with the water industry's official regulator on environmental issues, the Environment Agency, and increasingly favoured by local authorities, is SUDS (Sustainable Drainage Systems). It's hard to make this sound sexy to anyone but an environmentally aroused drainage engineer, but what it offers is a much more organic, flexible and attractive solution to the dispersal of groundwater than the traditional, Heath-Robinsonian apparatus of buried culverts and vulnerable pipework. This way, surface drainage becomes a feature in the landscape, and the natural water cycle – or something very like it – is part of the process. There is a variety of techniques. Permeable pavements can allow rainwater to soak directly into the subsoil, exactly as nature intended. Water may be collected in basins, or 'swales', whence it disperses into subsoil, pond or wetland; or it may flow into stone-filled trenches or drains for gradual filtration into the ground. Swales and filter drains can feed ponds and wetlands, which make their own visual and ecological contributions to the landscape – a refinement that delights environmentalists as much as it excites the cynics. Ponds, like canals, are magnets for the shopping trolley tendency, and hard experience shows what happens when they are neglected.

When unfiltered surface water does reach a sewer, it is no longer just 'rain'. It has become yet another liquid pollutant, bearing everything it has gathered on its journey across the surface. There is oil, dust, grit, carbon, hydrocarbons, zinc, road salt, agro-chemicals – whatever is lying around, sooner or later will find itself caught up in the tide and heading downstream towards the works. If it is in an urban area, the surge will be swelled by the outpourings from factories, which potentially include pretty much every substance that man knows how to make or use. There are exceptions, of course – some things are so dangerous or corrosive that they are banned from the sewers entirely and have to be disposed of, however dangerously, by other means. Other substances may be flushed only in agreed amounts specified in the trade effluent consents that water companies issue to industrial users. These set limits both for the quantity and the quality of what factories can discharge, and anyone who ignores one can look forward to an early appointment with the magistrates. Theoretically this gives the water companies tight control over what they have to deal with. In reality, however, they are bound to compromise. Just as with housing, they are not part of the statutory planning process and have no right to determine what kinds of industry operate in or move into their catchments. They could impose prohibitive limits and bring even a major employer to its knees, but the risk of job losses makes such belligerence politically awkward, if not actually impossible. By and large, they have to take what comes.

As we have seen, what comes is a formidable onslaught of liquid and solid, organic and inorganic, predictable and unimaginable. At journey's end, the sewage first washes up against a coarse screen that arrests the progress of large objects such as skateboards, murder victims and (a particular seasonal hazard) Christmas trees. The flow then continues into tanks, where it stands while the grit settles. This is known in the trade as the preliminary treatment stage. The effluent is still raw sewage, with an average live content of 10m faecal coliform bacteria in every 100ml of water. From here typically it passes through a 6mm stainless steel grid that sifts out the sanitary towels and other stuff from the U-bend flotilla, and onward into open tanks where it is held for an average of around two hours. This is long enough for dense solids to sink to the bottom, and for fat and grease to float to the top. The bottom layer is drawn off and the top layer skimmed, to leave a liquid residue that has now completed the 'primary' treatment stage, with contamination reduced tenfold to 1m faecal coliforms per 100ml. At one time this was thought good enough to pump straight into estuary or sea, but since January 2002 discharges from larger populations (of 15,000 or more) have had to progress to at least 'secondary' stage – a requirement that will extend to smaller populations (2000-plus) from 31 December 2005. Elsewhere, in remote areas where discharges are small and public access to the water limited, the outflow need be subject only to 'appropriate treatment', which, according to the Environment Agency, is 'locally defined according to environmental impact'. Even now this may mean primary treatment, or screening only. In the vast majority of cases now, however, the sewage moves on to 'secondary' treatment in a biological reactor – either a closed tank or a filter bed – where, in effect, it is fed to a dense colony of super-voracious bacteria that fall upon the organic content like piranhas on a pork chop. A few hours is all it takes. After the bacteria have feasted, the liquid moves on to yet another settlement process, where the sated bacteria settle in a dark brown layer at the bottom of what is now a tankful of clear water. In most cases, this is as good as it gets. The bacteria from the digestion process are pumped away to join the sludge drawn off earlier, and the water, with faecal coliforms now reduced to 10,000 per 100ml, is ready to be discharged into a river or the sea. It is just one more step in the infinite journey. Much of the effluent itself will be abstracted from the rivers, purified and sent on yet another round in its perpetual cycle via water main and teapot back to the sewer.

Secondary treatment is thought inadequate only where the effluent is heading for coastal water near bathing beaches, or for a river in which the Environment Agency demands particular standards. In this instance it may go for tertiary treatment, usually by filtration through sand, or disinfection by ultraviolet light, which will kill 99 per cent of the remaining bacteria and viruses and cut the faecal coliform count to an all but undetectable 35 per 100ml. What this means in most cases is that the effluent will be significantly cleaner than the river into which it is pumped. No precise figures are available, but the Environment Agency believes that 'something over 75 per cent' of the flow from sewage works into rivers is now treated to tertiary level. As we shall see, the industry is not beyond criticism (indeed, it attracts critics like its base material attracts flies), but it is entitled to feel that its very ubiquity – the sheer take-it-for-grantedness with which we dump on it everything we find most foul – denies it the credit it deserves. The fact is that the treated effluent from a major city will be less offensive than untreated sewage from a small caravan park.

Purifying the liquid, however, is only part of the answer: it still leaves the problem of the extracted solids, or sludge. This, too, has to find a home and, ideally, a use. Ironically, the solution is the same that the much-mocked Edwin Chadwick suggested right back in the 1840s – spreading on farmers' fields – though if Chadwick himself were to take a handful of modern treated 'bio-solids' he would be unlikely to guess its provenance. It no more resembles what you flush down the lavatory than this page resembles a fir tree. It's squeezed, digested by yet more bacteria, dried, sometimes even pasteurized, and pressed into cake or granules which the industry's marketing men are now pleased to call 'the product'. The quantity each company manages to dispose of in this way depends very much on its catchment – rural areas obviously have a much bigger appetite for agricultural fertilizer than industrial conurbations – though some will transport it beyond their local boundaries. Some continue to treat it as a problem of waste disposal and are pleased for farmers to take it off their hands for free; others sell it for profit. Yorkshire Water, with its mixed catchment, manages to give away about half of its sludge in cake form. Anglian Water, in rich beet and barley country, sells pretty much all it can produce. The dry granules are bagged and delivered to farms just like any other kind of commercial fertilizer, while the more mud-like 'cake' is stacked in field corners much like an old-fashioned manure heap. It is the kind of thing that makes proponents of sustainable agriculture purr with pleasure. Having been processed to a standard of purity which, at least in granular form, you could safely if not enjoyably eat, it is much cleaner than animal manure but delivers the same benefits to the soil. Even better, the digestion process at the treatment works generates bio-gas that can be used to create heat and energy within the plant – a significant gain in an industry that is a notoriously heavy energy-user.

There was a time, not so long ago, when raw or untreated sewage sludge went straight on to the fields without processing. Some still does, but only on to non-food crops – willow and poplar grown for coppicing, hemp for fibre, miscanthus for biomass – and even that will be phased out by the end of 2005. It has been banned on food crops since 1999. The modern treated bio-solids are made to two different standards, defined in an agreement – known with admirable directness as the Safe Sludge Matrix – between the water industry and the British Retail Consortium, voice of the UK supermarkets. In due course the matrix may become enshrined in law, but in the meantime it is the retail consortium, not Defra, that is the most powerful force in the UK food market, and the supermarkets that call the shots. If Sainsbury, Tesco and the rest won't buy into something, then you can forget it – it won't happen.

The lower quality standard under the matrix is for 'conventionally treated' sludges which have been biologically, chemically or heat treated to remove at least 99 per cent of the pathogens (organisms capable of causing disease). The higher standard, for 'enhanced treated sludges', guarantees almost perfect sterility. It is free from salmonella, and 99.9999 per cent of the pathogens – what mathematicians call a '6 log reduction' – have been destroyed. Even here, though, supermarket customers are double-insulated against any risk of contact, however remote, with the pasteurized products of their own metabolisms. Conventional sludge must not be used on fruit at all. With vegetables, it must be applied at least twelve months before harvesting – an interval that increases to thirty months if the crop is a salad likely to be eaten raw. For enhanced sludge on any kind of crop, despite its near sterility, the harvest interval is still ten months.

Such fastidiousness would have struck Edwin Chadwick's

regional correspondents, with their steamy reports of putrescence and contagion, as frankly comical. Other corners of the ordural empire, however, they might find oddly if not reassuringly familiar. No one these days is allowed to build a privy over a watercourse, but this does not mean that raw sewage never puts in an appearance. There are, indeed, odd echoes of the Middle Ages, caused in cities by the very same conjunction of factors – access to alcohol combined with the absence of lavatories. Cutbacks in local government spending have meant that, in less than ten years from 1995, the number of public 'conveniences' in England has been cut by nearly 10 per cent. There is now one for every 80,000 people in the UK (two thirds of which are designated for men). It is not just a question of the overall number – the problem is made exponentially worse by early closing. When it comes to lavatories open at night, the drop since 1995 has been 50 per cent. In London alone, the result is two million pints of urine sprayed in the streets every year, to which may be added, according to Keep Britain Tidy, a daily total of 'six deposits of faeces' and an average weekend outpouring of 300 pools of vomit.

But it is in the rivers and around the coasts that rogue sewage exerts its most baleful influence, with fatal effects on wildlife and visceral challenge to bathers. Only a great white shark can empty a holiday beach faster than faeces. In 2002 the Environment Agency recorded more than 2,000 pollution incidents involving crude sewage, of which 168 caused serious damage. At the minor end of the scale, riverside pubs were strangely (or perhaps not so strangely) frequent offenders. Breweries, like many other businesses without mains drainage in rural areas, often have special 'discharge consents' from the Environment Agency, which allow them to run their own private sewage-treatment plants and to pipe the effluent directly into brooks, streams or rivers.

Pub landlords on the whole are better at maintaining optics and beer engines than they are at sewage treatment plants, and there is an historically comic association between alcoholic and lavatorial excess. The result of a system breakdown, when customers unknowingly ape the medieval corruptcrs of the Walbrook, is locally disgusting but generally minor. Fish may suffer, and a few portions of scampi may suddenly reappear when their consumers hit the smell, but usually no lasting harm is done to anything but the pub's good name (a visit to the magistrates is usually guaranteed).

When the offender is a public sewerage undertaking, however, the problem is of a different order. In the past there were literally hundreds of local sewerage systems, with each rural or urban district council running its own. They were days of notorious stinks, with villagers frequently obliged to go about with handkerchiefs over their noses, and the letters pages of local newspapers filled with complaints. It was not until 1975 that the systems were consolidated and handed over to the ten regional water authorities, eventually to be privatized in 1989. Thus began the great irony of the water industry – the linking together of extreme purity and literally unimaginable filth. For thirty years, the same companies that provide our tap water have simultaneously gathered our sewage. The same industry that boasts of the world-beating purity of its product is also one of the country's most persistent polluters, regularly topping the Environment Agency's prosecution league.

In 2002 the industry itself caused 150 serious pollution incidents – most of them involving raw or partially treated sewage – and nine of the ten major companies shared fines totalling more than £1m. Three of them – United Utilities (which supplies north-west England), Thames Water (London and the south-east) and Anglian Water (east) – were singled out by the Environment Agency as 'significant repeat offenders' that had 'failed to learn from previous convictions'. Between them, the sewerage and water industries accounted for 11.4 per cent of all recorded pollution incidents. In 2003 things got even worse. 'Serious' incidents recorded by the Environment Agency went up by 23 per cent to 185 – the worst figures since 1998.

Many of these in effect were grossed-up versions of the pub incidents. Equipment failed; warning systems did not work; there was human error. In August 2002, for example, storm debris caused a blockage at Southern Water's treatment works at Lyndhurst on the river Beaulieu. This diverted the flow into a storm tank that overflowed and, for three days, poured untreated sewage straight into the river, with devastating effects on fish. The water was littered with dead lampreys, bullheads, stone loach, eels and trout, and the Environment Agency found nothing left alive within four kilometres of the spill. In June of the following year, a valve failed at a waste-water works near Formby in Lancashire. This allowed sludge to accumulate and build up dangerous concentrations of ammonia in nine million litres of effluent poured into the river Alt. Although it happened at night when no one was there to sniff the danger, an automatic telemetry system should have raised the alarm. That it failed to do so was due to one very simple oversight – the vital ammonia sensor had not been connected to the system.

On Good Friday, 2003, a cracked discharge pipe allowed sewage to spill on to Millendreath beach and give an unintended aptness to the name of Looe Bay. In August of the same year two pumps failed at a pumping station near Bexhill, in East Sussex, causing sewage to flood into an overflow tank, which in turn disgorged on to Cooden Beach (where swimmers were in the sea) after an alarm failed to work. In December, sewage was left to leak for several days from a manhole into a stream near Liverton in South Devon. Incidents like these may not be the most dramatic examples of failure; but they are typical of a system whose infrastructure creaks with age and begs for renewal. Typical, too, were the levels of fine imposed by local courts – £30,000 for Southern Water at Lyndhurst, £5,000 for United Utilities at Formby, £8,000 for Valley Waste Management at Millendreath, £3,700 for Southern Water at Cooden Beach, £5,000 for South West Water at Liverton. The courts are a persistent irritation to the Environment Agency, which argues that the penalties they impose – an average fine of £8,744 per offence – are too low to be a deterrent. 'Courts *are* getting tougher on environmental offenders,' says its chief executive Barbara Young, 'but fines are still small change for big business.' Almost always they are very much less than the cost of putting right any fault in the system – an imbalance which the agency has been lobbying the Lord Chancellor's Department and magistrates' courts to put right.

It also argues that companies should do more to ensure the efficiency of their early warning systems. 'They wait for the public to ring and tell them there are dead fish in the river,' says the EA's head of environmental quality, Dr Andrew Skinner. Even when mechanisms are in place, he says, they are often ineffective. 'Many companies have fancy alarm systems but haven't trained their staff to understand what the red light means.'

The industry for its part complains that it is held up like some shady, stink-in-the-hole bogeyman and pilloried for the sins of others. Having no part in the planning process, and being limited by Ofwat in the amount it can spend on renewing and extending its networks, it too often finds itself at the mercy of circumstances over which it has no control. 'New users can cause overload,' says Anthony Harrington, Science and Regulation Manager at Yorkshire Water. 'Overload can cause flooding, and flooding can cause pollution. Then we can find ourselves in front of the magistrates for incidents that are not our fault.' Worst of all are the customers themselves. The UK's biggest water company, Thames, insists that many of the pollution incidents for which the Environment Agency castigates it are the fault of others. More than half are sewer blockages caused by grease and fat poured down domestic sinks. DIY is another curse, with many amateur (and some professional) plumbers failing to recognize the difference between surface- and foul-water drains. In 2002 alone, Thames engineers identified 700 wrongly connected drains pouring sewage and washing-up water into channels designated for rainwater. Forty different streams were polluted as a result. In 2003, the Environment Agency asked Thames to investigate the contamination by sewage of Strawberry Vale Brook in St Pancras Cemetery, which was supposed to receive only surface water. As a result of detective work involving wire traps, CCTV cameras and coloured dyes, the problem was traced back to 183 homes in which a total of 191 appliances, including washing machines and lavatories, had been wrongly connected to rainwater instead of foul-water drains. In a further attempt to stem the flow of nuisance, and with commendable optimism, the water companies together launched a nationwide 'bag it and bin it' campaign, urging customers to flush down the lavatory only human waste and toilet paper (but not too much of it, and not the moist, extra-strong type). Everything else, including pet faeces, should either go out with the garbage or be returned to an appropriate collection point. And it's not just in the lavatory that habits need to change. The companies would like people to police their plugholes more effectively too: no more oil and fat to clog the underground arteries, please, and nothing whatever down manholes.

Even where systems do work properly, there are wider policy issues. Compliance with the law need not mean absence of pollution – it is all a matter of degree. The idea of 100 per cent purity in treated effluent is a myth: you can get rid of most of the viruses, most of the bacteria, but you can't get rid of them all. The question is: for the sake of public health, and for the safety of the environment, how 'pure' does the discharge need to be? In England and Wales, the body that sets the standards is the Environment Agency; north of the border it is the Scottish Environment Protection Agency (SEPA). The limits they set are so similar that, to keep things simple, we can take the Environment Agency as representative of both. Its remit is wide. The 'controlled waters' over which it has juris-diction include virtually all inland fresh water and groundwater, plus all tidal and coastal waters for three nautical miles out to sea. The only exceptions are self-contained reservoirs and ponds that do not connect with any other watercourses.

It is illegal for anyone to discharge into controlled waters any poisonous, noxious or polluting matter, any solid matter, or any trade or sewage effluent without a licence, known as a 'discharge consent', from the Environment Agency. Such consents are issued not only to water companies discharging through their treatment works but (as in the case of pubs) also to some factories and smaller businesses which – either because they do not have access to a public sewer, or because they produce more effluent than the sewer could manage – are allowed to operate treatment works of their own. Altogether some 100,000 discharge consents are currently in force, of which around 1,500 are classified as 'significant', meaning that they produce more than five cubic metres of effluent a day. Even sewage treatment works are of limited size. Only some 1,600 of them serve populations of more than 2,000, and of these only 700 cater for more than 10,000. Every one of them, large or small, has to process its waste to a prescribed standard, and the standard will vary according to the kind of water the effluent is destined for.

The quality of the discharge is controlled in one of two ways – by 'numeric' or 'descriptive' consents. Numeric consents are applied to outflows, including sewage treatment works, that have the greatest potential for harm. They impose maximum limits on the concentrations of particular substances, or groups of substances, in the discharged effluent, which has to comply with EC directives and international conventions. The numeric consent for a sewage works may include limits for a range of different substances, but the three most important categories are suspended solids, biochemical oxygen demand (usually abbreviated to BOD) and ammonia. Suspended solids include most obviously small particles of faecal matter, bearing their biological warheads of bacteria and viruses. Apart from their obvious health risk and offensive nature – in large concentrations they smell like what they are – they are immensely damaging to the freshwater environment. Suspended solids kill fish by clogging their gills, and other forms of life by blocking light from the water and blanketing the streambed. BOD is the amount of oxygen taken up by bacteria – if it is too high, fish in the water will suffocate. Ammonia is a normal by-product of organic decay which, in high concentrations, interferes with gill function and is lethal to fish.

When the discharge is from non-sewage sources, the terms are absolute. The Environment Agency demands full compliance at all times, without exception. Because water companies have less control over what their customers send through the sewers, some of them do have a little more room to manoeuvre. Although there is still an absolute maximum which must never be exceeded, they have the benefit of a 5 per cent foul-up rule. No offence is committed if they stay within normal limits for 95 per cent of a rolling twelve-month period. All numeric consents are regularly monitored by the Environment Agency, which samples and analyses water from the outfalls.

Descriptive consents are much less precise. They apply only to small operators whose discharges are of such low volume and toxicity that, in the agency's opinion, the risks are slight. The terms of the consent will cover the type of treatment plant to be installed, and may even prescribe the weather conditions under which it can be used – during storms, for example, when the dispersal of pollutants will be most rapid. Some 50,000 such consents are currently in force, of which the Environment Agency routinely inspects around 2,000 (although it does not usually sample the water).

We already know what can happen when consents are

breached – fouled beaches and bathing waters, disgusting smells, dead fish. But, according to the Environment Agency, non-compliance is not the problem it was. In 1964–5, for example, nearly 60 per cent of sewage treatment works breached their consents. In 2002 the failure rate was less than 5 per cent. On the face of it, we should all be taking deep, appreciative sniffs of the flower-scented air, and gambolling like baby otters in crystalline waters that teem with every species of aquatic life known to science. Environmental groups, which have kept up an insistent clamour for rivers and coasts to be cleaned up, should be cracking open a few bottles of well-chilled spring water and reflecting on a job well done.

Why, then, are they not? Are they such curmudgeons that they can't give credit for progress? Should we wonder, even, at their motives? Do they not depend for their very existence on having something to complain about? After all, what is their end product but indignation – a product which, like treated effluent, has to have somewhere to go. No matter how good things get, they will never be satisfied. Moan, moan, moan. The problem is that streams, rivers, estuaries and oceans mean different things to different people. A sewerage engineer sees a convenient medium for the disposal of ordure, and measures his success against a numerical standard set by law-makers. Anglers, sailors and swimmers see a leisure facility, and measure their success by how much pleasure it gives. Naturalists see a degraded environment, and measure their success by how many indicators of purity – crayfish, freshwater mussels, wild brown trout – they can nurse back into it. Law-makers see an eel's nest of colliding imperatives, and measure their success by criteria they invent for themselves. How good is good enough? Who knows?

Every day the sea around the UK receives some 300m gallons of treated effluent from sewage works. It sounds a lot, and it is. But in the marine environment, the word 'sewage' historically has borne almost as many interpretations as 'stew' might in a kitchen. It all depends what you put into it – or, perhaps more appositely, what you take out. For many years, following the failure of scientists to establish any link between polluted bathing water and the polio outbreak of the 1950s, it was imagined that there was no risk from swimming in any seawater that was not actually 'aesthetically revolting' – i.e., so thick with faeces that your stomach would heave at the very idea of contact with it. If you could enter the water without being sick, then you could splash in safety for as long as you liked. It took until 1976 for a more scientific approach to be agreed, when the EC Bathing Water Directive set statutory limits for faecal coliforms (intestinal bacteria, of which *Escherichia coli* is by far the most common) in 'designated bathing waters' – of which by 1979 there were a mere twenty-seven in England and Wales, not including even Blackpool or Brighton. Government lethargy drew a critical broadside from the Royal Commission on Environmental Pollution in 1984, but another five years passed before 333 more beaches were added to the list, which has subsequently grown to 483. Since 1998, nine inland lakes and rivers have also been designated. During the 2003 'bathing season' (officially 15 May to 30 September), 98.8 per cent of these met the EC standard and only three had to be temporarily closed for reasons of safety. In 2004, according to the Environment Agency, the pass rate nudged up to 99 per cent. This is consistent with a rapid period of sustained improvement from 1988–90, when the average compliance rate was only 57 per cent, to 2000–3, when it reached 95 per cent. Over the same period, bathing waters consistently falling below the standard went down from 13 per cent to zero. Part of the explanation may be that discharging sewage sludge anywhere into the sea was made illegal in 1998, which significantly reduced the likelihood of breast-stroking into what one commentator described as 'close encounters of the turd kind'. Another factor may be the £600m that the water industry calculates it has spent on improving the quality of its effluent.

In 2004, 375 beaches were given UK Seaside Awards by ENCAMS (Environmental Campaigns, formerly the Tidy Britain Group), and 122 resorts were given the top international Blue Flag award by FEE (Foundation for Environmental Education), to which ENCAMS belongs. This comfortably exceeded the previous record of 105, achieved in 2003, which itself was almost double the total for 2000 – another sure indicator of forward momentum. The problem here is that FEE, which draws its membership mainly from Europe but also from the Caribbean and South Africa, focuses its attention on marinas and resorts. To earn a Blue Flag, a beach must not only have clean bathing water and sand free from litter and dog-fouling – it must also offer a wide range of other, quite sophisticated facilities including lavatories, lifeguards, first-aid points and disabled access, which are not the kinds of thing you'll find, say, around the marshy coasts of the Wash or in a Cornish cove. ENCAMS has taken issue with its fellow members of FEE but has so far failed to persuade them to introduce a Blue Flag equivalent for unspoilt rural or isolated beaches where the very absence of infrastructure is an essential part of their character and appeal. ENCAMS's own Seaside Awards do include rural beaches, but still require much the same kinds of facilities as Blue Flag (though water standards are much lower). Beyond these, the only other designations are for the 483 statutorily recognized 'designated bathing waters', which are required to meet only minimum European water quality standards. Those which are not also Blue Flag or Seaside Award winners are not flagged in any way, so visitors – particularly on more isolated beaches – generally have no way of knowing how clean the water is.

The most persistent critic of UK policy is the pressure group Surfers Against Sewage (SAS), which accuses the government of failing to take water quality seriously. Its chief complaint is that by adopting only the minimum, 'mandatory' limits specified by the European Bathing Water Directive, rather than the much tougher 'guideline' ones, the UK applies the lowest standard it can get away with (though it must be said that in 2004, according to the Environment Agency, 80 per cent of bathing waters did meet the higher standard). The minimum standard is 2,000 faecal and 10,000 total coliforms per 100ml of water; the guideline standard 100 faecal and 500 total coliforms, and 100 faecal streptococci per 100ml. Only when water exceeds the mandatory coliform limit does the Environment Agency test for enteroviruses or salmonellae (and even then it does not act until the following year). And of course the standards apply only to 'designated bathing waters' anyway.

An even more basic question is whether the standards themselves are a realistic test of hygiene. SAS argues that coliforms alone are a poor indicator of pollution by sewage. The problem is that, in comparison with other micro-organisms, they die very quickly in salt water. The viruses that come with them, and that actually make people ill, survive much longer – up to a hundred days in the case of Hepatitis A – so that the absence of faecal coliforms in itself is no guarantee of health. SAS keeps a medical database which, it says, shows that people who swim or even paddle in officially 'clean' seawater have a significantly higher risk of contracting a wide range of illnesses including ear, nose and throat infections, gastrointestinal infections, wound infections, eye and skin infections, respiratory disease and a range of viral diseases. You would think there was not much room for argument here. It is incontestably true that viruses and bacteria travelling what medical men call the 'faecal-oral route' are common causes of gastro-enteritis (a loose term that covers everything from mild inconvenience to the risk of death) and the much more dangerous acute febrile respiratory illness. If you duck your head in polluted seawater, then the final, oral part of the route is wide open to any micro-organism that brushes your lips.

As you might expect, the World Health Organization has some opinions on this. Its 'Guidelines for Safe Recreational Water Environments', published in 1998, relied substantially on research funded by various arms of the UK government – the Department of the Environment (now Defra), the Department of Health, the Welsh Office and the National Rivers Authority (now the Environment Agency) – between 1989 and 1992. The subjects were adult volunteers recruited locally at four designated beaches in England and Wales, all of which met the official minimum standard for bathing water. In approved scientific manner, volunteers were randomly assigned to swimming and non-swimming groups and questioned about their health record. Swimmers were asked to stay in the sea for at least ten minutes and to duck their heads at least three times. Researchers meanwhile took water samples at a range of different depths. The health of the two groups – 507 swimmers and 605 non-swimmers – was then monitored for three weeks. The scientists' conclusion, enshrined by WHO in its guidelines, was that illness could be caused at levels lower than the official standards. Most strikingly they showed that, in the particular case of gastro-enteritis, a statistically significant increase in risk occurred at a level of only 32 faecal streptococci per 100ml of water – significantly lower than the EU guideline of 100 streptococci per 100ml. The risk was specifically associated with water sampled at chest height – the precise level at which bathers were most likely to duck their heads.

WHO was convinced, and SAS apparently vindicated in its campaign for cleaner water. Up from the statistical swamp, however, arose yet another official body funded by the UK government – the Medical Research Council's Institute for Environment and Health, based at the University of Leicester. Its report, 'A Review of the Health Effects of Sea Bathing Water', published in 2000, was massively long, jargon-ridden and repetitious. Nevertheless, screened by the verbiage, a ruth-less hatchet job was being done. In the authors' opinion, almost no aspect of the earlier trials was satisfactory. The selection of volunteers, the duration and quality of the monitoring, and the statistical analysis of the results all came in for sustained and heavy criticism, much of it opaque to laymen but nevertheless compelling in its thoroughness and detail. The authors repeatedly asserted their astonishment that the original report 'was regarded as suitable for publication in the form in which it appeared'. They did allow that some of the data might be useful, but not until they had been re-analysed by statisticians who knew the value of a good 'Monte Carlo simulation' or a 'Bayesian inference'.

As no such re-analysis has yet been made, it leaves the water little clearer than it was in 1998. The Bathing Water Directive is currently under review, and it may happen that the World Health Organization, UK government and all their squabbling scientists will line up behind the same set of figures. In the meantime it is left to the Environment Agency, and the sewerage companies it regulates, to decide what we swim in. The position of Surfers Against Sewage is clear and unequivocal: any sewage in the sea is too much. Research from the United States reinforces its view that European standards are set too high, and an authoritative study in the UK has shown that surfers – and, by extension, any others in regular contact with seawater – are three times more likely than other members of the public to contract Hepatitis A. For this reason, it argues, it shouldn't be left to wind and tide to disperse the stuff – we should do it ourselves by ensuring that *all* sewage effluent discharged into the sea has been through tertiary treatment and is as pure as we know how to make it. Although the unrefined, shit-where-you-like barbarities of the late twentieth century are going the way of the medieval dung heap, the ultimate state of bathroom-standard perfection may not be reached until such time as mankind finds some alternative to the daily voiding of its bowels.

It is one of the water companies' most plangent complaints that they are caught in a fork. On the one hand they are forced by their commercial regulator, Ofwat, to keep their prices down. On the other hand they are compelled by the Environment Agency, backed by an ever-expanding raft of

European law, to effectively perform the function of environmental contractors in cleaning up both their own act and the country's waterways. It was no surprise when Ofwat announced in October 2003 that the government, industry and public were facing 'hard choices'. The European Water Framework Directive sets stiff targets across a wide range of issues, from sludge treatment right down to the health of shellfish, that need to be met by 2012 or 2015, and there are fierce arguments about how much it will all cost, and how it will be paid for.

It exacerbates an old grievance. 'One of the truths about the water industry over the last fifteen years,' says a well-placed insider, 'is that a lot of money from customers has been spent on improving the quality of the environment – making sure that discharges from sewage treatment works are of the right quality and standard. What has been neglected is the replacement and renewal of the infrastructure itself. The result is that you're seeing burst mains, you're seeing collapsed sewers, you're seeing leaks. We're rushing around applying sticking-plaster remedies when we should really be discussing long-term renewal.'

Thames Water puts it bluntly. 'In previous price reviews,' its spokesman says, 'infrastructure has had a low priority. But now the emphasis has to change. We are seeking funding support [i.e., price increases] from Ofwat for an accelerated replacement programme.' Water UK (the trade association of privatized water companies) calculates that, to meet its improvement targets, bills would have to rise by an average of 31 per cent from the 2002 level, to £306 a year by 2010. The Environment Agency meanwhile insists that the cost of improving beaches, rivers and wildlife habitats would add to customers' bills 'the price of a can of fizzy drink each week'. At the same time its head of environmental quality, Andrew Skinner, welcomes the fact that the environmental standards set by the Water Framework Directive will cost money. For this reason, he says, they will continue to dominate the industry's spending priorities. Yet he is impatient of the companies' tendency to represent themselves as 'fall guys expected to set the world to rights when a lot of the issues are the responsibility of others, especially farmers'.

'We start parting company with Water UK when they say they have done all they have to do, and don't see why their customers should pay more for environmental improvements.' In 2002 the agency, having calculated the value of its can-of-drink policy, proposed a five-year action plan to upgrade the quality of 4,000 different sites, affecting the health of 6,500km of rivers and 2,000 sq km of coastal waters, still waters and wetlands. Wise in the ways of a calculating and cynical world, it did not argue solely on the basis of sentiment. Academics and economists testified that the benefits of cleaner rivers and wetlands to the leisure, tourism and other industries would be in the region of £5 billion to £8 billion over twenty-five years. With varying degrees of enthusiasm, the water companies did their costings and duly included all this in their draft business plans for 2005–10 which they submitted for examination and approval by the commercial regulator, Ofwat. What they wanted was an average 29 per cent above-inflation price rise spread over five years; what Ofwat seemed to be offering in its 'draft determination' in August 2004 was just 13 per cent. In the event, its announcement in early December of an 18.4 per cent five-year hike caught many people by surprise. In public at least, Water UK kept its counsel. It welcomed Ofwat's acceptance that more investment would be needed to repair and replace parts of the system, but regarded it nevertheless as a 'tough settlement'. 'Companies will be looking carefully at whether Ofwat has found the right balance between risk and return for them,' said its chief executive, Pamela Taylor. 'Above all, they want to be sure that the prices they can charge will allow them to supply the services people want and meet their environmental obligations.' If they felt they couldn't, then they might have to refer the issue to the Competition Commission.

Politically, the whole issue is a nest of vipers. Access to clean water is universally regarded – if not actually respected – as an inalienable human right, and its absence as a sure sign of degradation and hardship. Putting a price on it is like putting a price on life itself. When the water supply on South Africa's eastern seaboard was privatized in 1999, thousands of slum-dwellers were unable to afford the bills and had to find their water from other sources. Result: the twenty-first century announced its arrival with the worst cholera outbreak in the country's history: 250,000 people infected, 300 dead. As always, the announcement of increases in the UK provoked an outcry from representatives of the less well-off. 'The burden of water bills for people on fixed incomes,' said the National Consumer Council, '. . . is increasing out of all proportion to their ability to pay.' Meanwhile the Environment Agency complained that the environmental improvements made affordable by the spending plans were no more than the minimum required by European law. For governments, it presents a classic moral dilemma. If they let prices rise, they risk hostile headlines and political damage. If they keep them down, then it is some future administration that will inherit the electoral backlash.

In terms of its wider impact on the environment, the water industry faces a double whammy. On the one hand, its sewage disposal arm is a convicted polluter. On the other hand, its water supply arm is accused of degrading rivers, and of robbing wildlife of its home, by taking too much water. According to the Environment Agency, some 60 per cent of the total area of freshwater and wetland Sites of Special Scientific Interest (SSSIs) in England and Wales are in what it calls an 'unfavourable condition'. The thinking behind the twinning of the water and sewerage industries in 1975 was utterly cogent. As providers of tap water, the companies would have a vested interest in keeping river-bound effluent as clean as possible, for they themselves would have to bear the cost of processing and purifying it for recycling into the water supply. But, just as they have no direct control over what is discharged into their sewers, so – with the single exception of their own outfalls – they have no direct control over what seeps, leaks or is otherwise discharged, legally or not, into streams and rivers. In this they perceive two injustices. First: when it comes to pointing the finger and hauling someone off to court, they are a soft target. If they are fined more often than any other industry, then that is only to be expected. They handle more pollutants than any other industry, and it's in the nature of the business that every time something goes wrong, however minor, there is likely to be a breach of the law. Second: at their freshwater treatment plants they have to scrub out not just the pollution for which they are directly responsible but everyone else's as well. By and large, other polluters – the slow tricklers – get away with it and are never detected and pursued. This is particularly galling for the water companies when they reflect upon the £5 billion they have spent on improving their sewage treatment works and cleaning up their own mess.

Only when gross, one-off spillages are easily traceable to single sources does the Environment Agency take action. In December 2003, for example, it prosecuted a catering company in Leamington for polluting the river Leam with cream – a substance which, despite its benign association with strawberries, turns nasty when it hits a river, where it has the organic strength of raw sewage. A few days earlier the agency took to court a Devon brewing company that so badly contaminated the river Clyst with brewery sludge that agency officers had to dose the water with hydrogen peroxide. On 21 November of the same year, it successfully prosecuted a Sussex farmer who had emptied a small container of herbicide on to an area of grass, whence it had drained via a ditch into a pond, where it killed 200 fish. On 27 November, it was the turn of a Dorset farmer to face the magistrates, who fined him £500 with £2,800 costs for polluting the river Divelish with farm slurry. On 16 December, Thames Water itself was prosecuted for allowing diesel oil to escape into a stream at Crawley. On 13 November, a meat wholesaler was fined for a similar offence on the rivers Thet and Whittle in Norfolk.

All these prosecutions occurred during a period of just forty days at the end of 2003 (i.e., they were recent at the time I was preparing this part of the book). While they are not offences of particular seriousness, they do provide an illuminating snapshot of what the freshwater environment routinely has to suffer, and of what those responsible for the water supply have to cope with. They represent only a fraction of the problem. All of them, like sewage spills, are examples of 'point' pollution – specific incidents that are traceable to a particular source at a particular time. Offenders can be named, shamed and punished. Far more insidious, and much harder to prevent, is the 'diffuse' pollution that seeps quietly and persistently into the water from a multiplicity of untraceable and mostly unstoppable sources. Almost every kind of land use – urban, industrial, transport, forestry – has a contribution to make, but the overwhelming majority of diffuse water pollution comes from farming. Given that agriculture still occupies 76 per cent of the land surface of England and Wales, and that there are at least 175,000 farm holdings of widely varying size and type, this is not likely to raise too many eyebrows. (One might be more surprised, indeed, to learn that 'only' 50 per cent of the phosphates and 70 per cent of the nitrates in UK freshwaters come from agricultural sources.) Nitrate levels of over 30 milligrams per litre of water – the level at which it becomes a threat to wildlife – have been reached in 32 per cent of English rivers. Perversely, in the countryside, rain is not the universal cleanser of the poetic imagination. 'O thou lord of life,' wrote Gerard Manley Hopkins, 'send my roots rain.' Little can he have imagined how often such a prayer would be answered with a curse. Agricultural pollutants – pesticides, fertilizers, manures – reach watercourses by one of two routes: in surface run-off, or by leaching through the soil. Both processes are critically accelerated by rain.

The effects on water can be devastating. When river levels are low, it is like taunting a thirsty man with a cup of poison. Dissolved ammonia and pesticides are toxic, and ultimately fatal, to plants and animals. Nutrients (the most important of which is phosphorus) cause 'eutrophication', a state in which water becomes so over-enriched that it stimulates an excessive growth of algae. When this happens, the precariously balanced pyramid of aquatic life is knocked off its ecological tightrope. Submerged flowering plants are killed and the water and streambed become choked with algae, to the exclusion of almost all other forms of life. Algal blooms kill fish by depleting the oxygen in the water, and algae on the streambed blanket the sediment and throttle all the species that would normally live, feed or breed there. Such is the price we pay for the 'cheap' food from the subsidy-driven, wipe-down agricultural landscape where the 40ft spray boom is the wizard's wand whose magic must never fail. Wildlife has long been cleansed from these industrialized arable crop-zones, where a corncrake now has no more business than a rat in a Pot Noodle factory. As it is in the fields, so it is in the water.

Unfortunately it's not just ugly farming that produces ugly results. Picturesque, old-fashioned, free-range stock-rearing can be just as harmful. Where cattle have access to streams or rivers, they add to the burden of silt by trampling and eroding the banks, and pollute the water by urinating and defecating into it. Given the fanatical care with which human ordure is supposed to be handled, and the near-hysteria that greets any escape of it into the environment, the absence of public concern about its animal equivalent is deeply mysterious. Into the water with the ubiquitous cowpat go a tribe of unfriendly pathogens including the intestinal parasite cryptosporidium, which is directly linked to outbreaks of gastrointestinal disease in humans. At least eighty-one coastal bathing waters in England and Wales are believed to have failed the mandatory standards for faecal coliforms as a direct result of river water carrying farm, not human, manure into the sea. It is not the least of the water industry's complaints that screening for cryptosporidium alone can add £10m to the capital cost of a treatment works. Defra calculates that the total annual cost to the industry of treating drinking water for pesticides, nitrate, phosphate, micro-organisms and suspended solids, much of which comes from agriculture, is £225m – another factor that disguises the true cost of food in the shops.

Animals cause damage in other ways too. The rooting habits of outdoor-reared pigs, for example, cause serious soil erosion – though this may be nothing compared to the loss of soil from arable fields. According to Defra, the last twenty years have seen what it calls a 'marked increase' in the volume of soil washing off the land into watercourses. The figures are so unbelievable, I asked Defra to recheck them. Typical run-off rates are between 0.1 and 20 tonnes of soil per hectare a year, and in extreme cases may reach as much as 100 tonnes – equivalent to the body-weight of 1,432 eleven-stone farmers. This not only represents a damaging and ultimately unsustainable loss to the farmer, who will need more and more chemicals to compensate for the loss of fertile soil, but also it deals another hammer-blow to the streams and rivers. The silt clouds the water, blocking the light that plants need to grow, and effectively blinds animal species that hunt or avoid predators by sight. It also clogs the streambed, thus eliminating yet more plants as well as a wide range of fish and invertebrates that need access to clean gravel at various stages in their lives. Salmon and trout, which bury their eggs in it, are particularly badly affected. On top of all this, the soil brings with it a further load of pesticides, phosphorus, manure and other pollutants.

It is not just obscure ditches and local rivers that are suffering. Even Lake Windermere is vulnerable. A build-up of phosphorus since the 1960s increased algal growth to the point where oxygen concentrations in the deeper water were throttled down to zero and the Arctic charr began to suffer. Around the margins, too, the alga *Cladophora glomerata*, familiarly but unpopularly known as blanket weed, over-whelmed all other growth and displaced the native brown trout. According to the Environment Agency, without action to reduce phosphorus the entire future of life in the lake will be 'on a knife edge'. Given that the Lake District has been pro-posed as a World Heritage Site, and that Coniston Water is suffering much the same kinds of problem, the inability of even the UK's most poetic landscape to protect itself is a clear indicator of the true measure of the problem. Forty-six per cent of English lake SSSIs are now affected by nutrient enrichment, and English Nature reports that levels in rivers have increased by 200–300 per cent since the 1930s. In 2001, 55 per cent of English rivers had phosphorus levels high enough to cause eutrophication.

Other victims include the famous chalk streams of southern England. The Hampshire Avon has suffered such an onslaught of sedimentation and over-enrichment that its crystal waters have clouded over and the gravel beds – historic breeding grounds of trout and salmon – have become overlain with silt. As a result, its once-perfect ecosystem of water plants, insects and fish is being supplanted by algae. Between 1989 and 1999, hatchings of fourteen different species of aquatic fly – the essential food of trout that fly-fishermen mimic with their lures – declined on chalk streams by 52 per cent. If a showcase river such as the Hampshire Avon can go to the bad, even at a time when the Environment Agency is acknowledging major improvements in the overall chemical and biological quality of Britain's rivers, then what hope is there for the rest?

In the case of the Avon itself, and of its neighbours the Itchen, Test and Rother, the specific hope is that the agency will succeed in its 'Landcare' project – which sets national standards for soil erosion, soil structure and levels of organic matter – to persuade farmers to adopt 'best practice' and cut off diffuse pollution at source. Defra, too, has been showing increasing concern, and has been consulting widely on how the problem might be tackled. Much rests on long-term reform of the disastrous Common Agricultural Policy, with the emphasis of the subsidy system swinging from quantity to quality so that farmers will be rewarded for environmental improvements rather than the damaging chase for ever more colossal yields. From the water industry's point of view, the parallel pollution of groundwater – water held in the soil, aquifers or permeable rocks, accessed by wells or boreholes – is just as much of a problem. In November 2004, UK Water Industry Research (UKWIR), the scientific arm of Water UK, reported a serious deterioration in the quality of groundwater over the previous thirty years. In 1975, it said, most of it required only minimal processing. Now nearly half of it needs major treatment, and with further deterioration over twenty years or so this would probably increase to 100 per cent. The capital cost of extra treatment since 1975 (part of it, admittedly, accounted for by tighter drinking water standards) stood at £658m, and the British Geological Survey was predicting future costs of between £73m and £180m for each five-year period until 2030.

The infernal twin of pollution is over-abstraction – the removal from a river or borehole of so much water that the entire character of a waterway or wetland is altered. For more than a decade since the Rio Earth Summit in 1992, the rallying cry in field, fen and forest has been 'biodiversity' – a grand, portmanteau word that embraces everything alive. This, and nothing less, is what we are now called upon to preserve. Every species, every ecosystem, every habitat. 'Biodiversity' is not just lions, tigers and whales. It's mould, weevils, intestinal parasites; nature with its trousers down. It means recognizing that, in the over-arching scheme of things, it is not possible to know exactly how each plant or animal contributes to the wellbeing of the whole, and that each has a value of its own. Crucially it means rescuing and preserving in its natural habitat any species facing extinction or grievous decline. In many, if not most, cases this is less a matter of protecting the plant or animal itself than of preserving the habitat in which it lives.

In Britain the engine of recovery, launched in 1994, is the UK Biodiversity Action Plan (BAP), which now embraces strategies for 47 different habitats and 391 species. No fewer than 106 of the priority species are associated with water and wetland habitats which, between them, hold more than a third of all the UK's flowering plants. Species in peril include the water vole, natterjack toad, native white-clawed crayfish, wild salmon, pollan, freshwater pearl mussel, depressed river mussel, medicinal leech and a large number of plants and insects, including the lesser silver water beetle, the Pashford pot beetle and the diving beetle *Bidessus unistriatus*. The Royal Society for the Protection of Birds (RSPB), which has been unhappily watching waders disappear with the wetlands that kept them alive, is a persistent critic of over-abstraction, pollution and other corrupters of the wild and the wet. Even since the early 1980s, it says, lapwing have declined by 47 per cent, curlew by 31 per cent, redshank by 40 per cent and snipe by 65 per cent. Half the wetlands outside protected areas in 2002 had no breeding waders at all. In an island as heavily populated, car-fixated and densely cultivated as our own, not even a migrating goose would expect to fly into a landscape of primeval wilderness. Even so, the speed and extent of habitat loss is a poor advertisement for our scale of values. The scarcer and more fragile an environment becomes, the more ruthlessly we strive to erase it from the map. On the national balance sheet, which leaves unpriced anything that cannot be taken to market, these are rubbish landscapes, devoid of profit and, hence, of purpose.

Undrained fen, lowland raised bog, reedbed and coastal and floodplain grazing marsh, like chalk rivers and mesotrophic lakes, are all priority habitats under BAP. We protect them or we lose them. The Anglo-Saxons on their fifth-century rampage from the east coast westward towards Wales are reckoned to have swung south through Hertfordshire to avoid the boot-sucking swamps of the Lincolnshire, Norfolk and Cambridgeshire fens. In the seventeenth century, Charles I and Oliver Cromwell urged their armies around a country that still soaked its feet in 3,380 square kilometres of undrained fen. By 1934 this had shrunk to 100 sq km, and by 1984 to just ten. The very rarity of such places now ensures that any survivor will be a designated SSSI or nature reserve, though not even these are immune from assault. The RSPB's own reserve at Strumpshaw Fen in Norfolk – a rare haven for bitterns and swallowtail butterflies – is degraded by loss of water through a neighbouring borehole. It's much the same at Fowlmere in Cambridgeshire, a reserve that offers sightings of water vole, southern marsh orchid and white-clawed crayfish. Its vital spring flows are curtailed by no fewer than thirty-six groundwater abstractions within five kilometres of the site. Other wetland habitats are under just as much pressure. The area of undisturbed lowland raised bog has shrunk by 94 per cent since the beginning of the twentieth century. Only 200 hectares now remain. Reedbed is down to 5,000 hectares. Even in the Norfolk and Suffolk Broads, 50 per cent has been lost since the 1930s. 'Feather-footed through the plashy fen passes the questing vole'? Evelyn Waugh's wincingly accurate parody of nature writers' prose now hits a different nerve. Altogether, the Environment Agency calculates that some 500 rivers, wetlands and lakes are at risk from abstraction.

The greatest frustration for those watching fens being sucked dry and once-great rivers reduced to trickles is the volume of water that goes to waste. The amount lost to bursts and leaks may have gone down year by year, but the industry admits that there has been no reduction in the total number of failures – it has just got better at finding and repairing them. Even so, the figures are mind-numbing, and leakages will go on multiplying as the industry's ancient pipework surrenders to age. *Every day* throughout 2002–3, an average of 3,600 million litres of water – enough to fill 1,440 Olympic-size swimming pools – gushed uselessly out of the system, 6 per cent more than the previous year, and in some areas 20 per cent of the entire supply.

Britain's water and sewerage systems deserve a place in any museum of engineering excellence. But a museum is where they now belong. It is a testament to bygone genius that parts of Bath's sewerage system – still functioning – were built by the Romans, and that many of the cast-iron trunk mains throughout the country were laid down by the Victorians. You do not need the brain of Brunel to work out that, after a century-and-a-half in the ground, even top-quality Victorian ironwork is deep into its operational twilight. In the Thames region, half the network is more than a hundred years old, and a third of it older than 150, much of it battered by the weight of traffic from above and the heave of London's clay beneath. All this is hostage to the Ofwat price-juggling exercise, but, as we have seen, the way ahead is far from clear. Costly vision, or cut-price pragmatism? Who knows? Not until the cost of repairing old pipes exceeds the cost of replacing them will progress be guaranteed. At the current rate of renewal, some of the Victorian pipework will have to serve until the middle of the twenty-third century.

The thread by which hope now hangs is the Water Framework Directive, a great sprawl of European regulation that gives national governments until 2015 to protect and enhance the quality of their waters, and to protect their surviving wetlands. All will depend on how, after the inevitable round of consultations with 'stakeholders' (or 'the usual suspects' as a Defra official put it), the UK government interprets and warms to the task; and how well the two regulators, Ofwat and the Environment Agency, resolve the often conflicting demands of price and the environment. For its part, the agency is as clear as mountain springwater on its four priorities: to protect habitats from pollution and over-abstraction; to stop pollution from sewerage overflows; to banish phosphate from lakes and rivers; and to stop leakage.

For the rest of us, it is a matter of each accepting our grain of responsibility. Of binning and bagging, not flushing solid objects down the lavatory. Of keeping fats out of the sink. Of using no more water than we need. There is no pain in any of this; no loss of convenience or damage to the quality of our lives. Installing a water meter, for example, will ensure you pay only for what you use; if you waste less, you pay less. And wasting less is easy. If you don't have a low-flush lavatory, convert your existing one by putting a sealed, water-filled plastic bottle in the cistern. Take showers rather than baths – that way you'll use only half the water and energy. Use washing machine or dishwasher only when it is full. Turn off the tap when you clean your teeth – it will save eight litres a minute. And fix leaking taps – a once-per-second drip-rate costs five litres an hour. In the garden, install a water butt, grow drought-resistant plants that need less water, and use a watering can instead of a hose.

Well, yes, I know, you can hear the yawns. But if we don't take seriously our responsibility to conserve water, then we don't take seriously our responsibility to the environment, and we disqualify ourselves from any right to criticize others. If fly-tippers are selfishly degrading the landscape, then who are we to complain if we duplicate their assaults from the secret havens of our kitchens and bathrooms? 'Environment' has become a word like 'heritage', damaged by association with worthiness. But it is not an abstract concept privately owned by sandalistas. The environment is the world in which we live. If we want to share it with other creatures – to have fish in the rivers, butterflies and birds in the sky, clean water to swim in as well as to drink, an aquatic landscape that is a joy to the eye rather than an offence to the nose – then we'll have to recognize that the tap turns both ways.

CHAPTER THREE

Trashing the Land

S
T JAMES'S PARK, LONDON, A FEW HUNDRED YARDS NORTH
of Buckingham Palace near Queen Anne's Gate. A small
crowd gathers, making little whoops of excitement.
Cameras flash. More people come running, ignoring the rain.
The object of their interest obliges with a fractional, freeze-
frame moment of perfect stillness, then flips through 180
degrees and shoots skywards. Twenty pairs of eyes follow it up
the tree. *Squirrel!*

It's a bog-standard grey, *Sciurus carolinensis*, unpopularly
known as tree-rat by people who mourn – though they may
never have seen – the native red squirrel of nursery romance,
Sciurus vulgaris. Nevertheless, it *is* a squirrel. An alert, wild
creature in the middle of a great city. A curiosity, anachronism
even, yet also a kind of comfort food for the eye. There is a
feeling of reconnection, a faint stirring of atavism that is more
than just yearning for a bygone age when, in some forgotten
way, man and nature pulled together. Somehow, though lacking much understanding of what its life is actually like, we
envy the squirrel its sense of belonging.

71

It is a sense that we have stripped from ourselves as completely as we have stripped the identity from lowland landscapes, villages and towns. Where distinctiveness survives, it is in relict form – a conservation village with stone or timbered cottages; a nature reserve; a squirrel in a park. The sense of waste eats into us like acid into limestone. We connect; we don't connect. Atavism lights a spark that dies in the cold hearth of nostalgia. We, the generation of the bulldozer and the wrecking ball, are the first people on earth to mourn the passing of a world we never knew. And even as we feel the pain of it, we seem unable to hold ourselves back.

No people in history have understood less about their landscape than the post-war British, or done more to smash what they claim to love. Most now could not tell a dunnock from a dustbin lid. Even country people, born and bred, are now bewildered by the things they see happening around them. Modern agriculture is a spray-on product that comes out of a packet, not out of the accumulated experience of generations of landsmen. The hedges and trees have gone, so nobody knows their names any more. Herbicides and insecticides have cut off the food chain at its roots, obliterating wildlife; and we have already seen what's happened to the rivers. Towns and villages which, in the post-war building booms, could have elevated the human spirit, instead have crushed it. Rural England is where urban England now dumps its rubbish – here it tips everything from garbage in landfills to fridges in ponds, broken cars and surplus people.

It could not be said that no one saw it coming. On 27 June 1939, radio listeners in the West Country heard a talk by

John Betjeman on the subject of country towns. Sixty-six years later, the points he raised seem commonplace. Don't we all know our best towns have been laid waste? That architects, planners and developers let off the leash by indolent, stupid or self-interested councillors have spread the ugliest rash of buildings to have appeared in Britain since mud first met thatch? That trying to replace what we've thrown away is like trying to make water flow uphill, or time run backwards? In the Thirties, however, the likes of Betjeman and Clough Williams-Ellis were the true prophets of their age. In their opposition to certain kinds of change, they might have looked behind the times. In anticipating the results, however, they were a good thirty years ahead of the game. The National Grid was new. So were the lower middle classes with their white collars fresh out of the bag, and their mortgage-fed appetite for self-improvement. So were the suburbs they lived in. So, in terms of mass ownership, was that ultimate symbol of twentieth-century advancement, the motor car.

Williams-Ellis and Betjeman saw only too clearly where it would lead. Pylons stamping through the fields. Concrete lamp standards marching through the towns. Fine old buildings knocked down to make way for traffic. Garish, over-lit shopfronts and huge advertising hoardings. A flood tide of urban sprawl that was not only ugly in its own right but repudiated any sense of relationship with the landscape on which it was inflicted. The loss of grain and texture in landscape and town alike. The rejection of human scale. The victory of the highways engineer *über alles*.

Betjeman's radio talk was a rallying cry. It was clear to him what men and women who loved England should do: 'Get on the local council, force an election . . . Save England from bad local government; save our old towns, our stately Georgian streets, build worthy new towns in the right place.'

Eleven years earlier, in *England and the Octopus*, the book that raised the standard for the new Council for the Preservation of Rural England (CPRE – now the Campaign to Protect Rural England), the visionary architect Clough Williams-Ellis had written with such choler that he all but boiled the ink. Our great- and great-great-grandfathers, he said, 'had a general sense of order and beauty, and realised that appearances were not "mere" but had profound if subconscious reactions on people's lives'. Yet the towns and villages they left us had already become part of the national myth – the lost fabric of the historical continuum, in whose shadow we would substitute nostalgia for optimism and find ourselves grasping at squirrels.

'In grouped architecture and visual country amenity,' Williams-Ellis wrote, 'we have evidence enough to show how calamitously we have fallen below all other ages whatever in the very things that distinguish the civilised from the savage. Can we believe that we, the English people, have thus fallen from grace for ever, that never again will England be an island of unsmirched country and ordered towns?'

This, remember, was 1928. What, in the coming decades, would become of Williams-Ellis's 'gospel of beauty in the common setting of our daily life'? The most inexcusable of our failings has been in the quality of what we have built. New development, wherever it appears, is almost always opposed by the people who will have to wake up and look at it. In part this is due to the sheer volume of concrete implicit in successive governments' housing projections. Even if John Prescott succeeds in shifting 60 per cent of new housing into towns, it could still leave 260 square miles' worth chasing greenfield sites in the countryside. But quantity is not the only issue here. People object to new development not because they think building is unnecessary or even misplaced (though it often is). What drives them into opposition is the sheer awfulness of what is built.

Modern domestic architecture in Britain is a miracle of perversity. In a country with such a rich tradition of vernacular building and grand design, how could we have sunk to such a nadir? Given that variations in form, scale, materials, colour and layout are so nearly infinite, how could we have settled for such dull and dispiriting uniformity? Once, not so long ago, houses spoke, like their owners, with local accents. Granite and slate; timber-frame; cob and thatch; clunch; brick-and-tile; flint; Cotswold stone – each material echoed the landscape from which it sprang, and gave the buildings their colour and form. You could have parachuted blindfold into any county in England and known where you were simply by looking at the houses.

Now we have homes for the hamburger age. The size may change, but the recipe is the same in Cumbria as it is in Cornwall. In place of Williams-Ellis's 'grouped architecture' we have ranks of indistinguishable, computer-generated housing products facing each other across cul-de-sacs wider than most village high streets. The builders argue that not all the houses are identical, and of course they are right. Out of the dressing-up box comes a pick-and-mix of architectural flim-flam – bottle-glass, leaded lights, 'Georgian' porticos, Tudorbethan timbering, gabled dormers. All to be stuck on to this or that 'dream home' without regard for balance, proportion, architectural cogency or the house's relationship with its neighbours. Architectural critics dismiss this kind of thing as 'developers' baroque', but the builders themselves are nearer the mark. Their private term for such adornments is 'gob-ons'. Planners hate them, or say they do, and yet for decades all over the country, and in full knowledge of what they were doing, councillors have been sticking up their hands in favour. 'Get on the local council,' Betjeman urged, all those years ago. What he couldn't have foreseen was that our representatives would use their votes not to 'build worthy new towns in the right place', but to turn their districts into architectural junkyards. What on earth made them do it?

The answer lies in the very system that most people thought was there to protect them – the planning process itself. Towns voted in favour of self-mutilation not because they liked what they saw but because they had no alternative. Design standards, already low, were decisively shot to pieces in the 1980s when the then Department of the Environment issued one of the most notorious decrees in the entire history of town and country planning – Circular 22/80. What it said was that local authorities had no business worrying about what new houses looked like. They could refuse permission on 'planning grounds' – i.e., the suitability of a site for development – but not on grounds of ugliness. The sole arbiters of taste would be market forces, just as they would be in every other compartment of late twentieth-century life. Reality TV shows couldn't be trash if people watched them. Hamburgers couldn't be junk if people ate them. Houses couldn't be rubbish if people lived in them. Thenceforward, any planner dabbling in the stuff of other people's taste could expect to have his nose bitten off. A builder denied permission could appeal to the DoE's planning inspectorate in the near certainty that both judgment and costs would go his way. Just to make sure everyone knew where they stood, the DoE reinforced its edict in a Planning Policy Guidance note known as PPG1.

It was as if a gun had been fired to set off an ugly building race. One of the biggest private building booms Britain has ever seen could have added something positive to the landscape – it might even have developed into a new, classic age of domestic architecture. Instead, we got what we all can see. As the tide of ticky-tacky rose, not even a market-fixated government could sustain the pretence that it liked what it saw. The Tory party's roots were in the countryside, and voters objecting to 'executive-style' excrescences did not enjoy being dismissed as Nimbys. How could it be believed that those who had to live next door to an eyesore were less entitled to an opinion than those who intended to profit by it?

In 1991 the then Secretary of State for the Environment, Chris Patten, ordered a rethink. To the guidance note PPG1 was attached an addendum known as Annex A. Struggling with conflicting instincts, it delivered a Janus-faced compromise that pointed in both directions at once. 'Planning authorities,' it said on the one hand, 'should reject obviously poor designs which are out of scale or character with their surroundings.' 'Aesthetic judgments,' it said on the other, 'are to some extent subjective and authorities should not impose their taste . . . simply because it is superior.'

No word in the lexicon of government carries more negative freight than 'subjective'. Planners would rather have their eyelids glued open for a fortnight in Milton Keynes than admit that there was such a thing as fundamentally 'good' or 'bad' design. Hackles rise at the very sound of words such as 'beautiful' or 'ugly', as if the failure of beauty to quantify itself is a full and adequate justification for its extirpation from the language.

In February 1997, PPG1 was changed yet again. The 'poor designs' which should be rejected, it said, could include those which were 'inappropriate to their context . . . clearly out of scale or incompatible with their surroundings'. As ever, local authorities were warned against interfering with other people's taste, but this time there was a caveat – 'except where such matters have a significant effect on the character or quality of the area, including neighbouring buildings'. It also accepted that it was 'proper to seek to promote or reinforce local distinctiveness'.

It looked good but it meant little. One is reminded of the paradox of the battery hen. Welfare campaigners object to its confinement and argue for a more free-ranging 'natural' environment. The snag is that a battery hen is no more like its remote wild ancestor, the red jungle fowl, than the Chancellor of the Exchequer is like a Palaeolithic hunter-gatherer. Its nature has changed. And so has the nature of Britain's towns and villages. Gob-ons apart, traditional building forms and local building materials have gone the way of the chicken's powers of flight: they are no longer in their make-up. It is simply too late now to start talking about 'local distinctiveness'. The reality is that the 'character and quality' of most places has already been altered by the introduction of national house styles, and compatibility just means more of the same.

Housebuilders heard what the legislators were saying, and picked up their cue. If they weren't building what people wanted, they said, then the houses would be standing empty. The fact is, of course, that people can buy only what they are offered. Where buyers do have a choice, market forces tell a different story. Homes are one of the few commodities in which second-hand values are usually higher than new. Almost everywhere in England, brand new semis and detached houses fetch lower prices than those built before the Second World War, and much less – at least 20 per cent – than those built before the First. The only homes that sell for less than their new-built equivalents are the architectural dregs of the 1970s and '80s.

Local authorities know all this, but – like robbery victims handing over their wallets to avoid assault – many have lost the will to resist. They close their eyes, raise their hands and think of PPG1. Others show more awareness, if not always the full courage of their convictions, by publishing local design guides. These stop short of laying down specific house-types for builders to copy ('prescriptive' to a planner is the next worst word after 'subjective'), but they do give examples of local vernacular style and materials sufficient to help a builder understand the local tradition and create designs that at least are compatible with it.

It is talk like this that makes accountants reach for the liver salts. Houses are like cars, they say. You can't jack up the design without also jacking up the price. But the argument is as phoney as a Tudorbethan gable. Housing associations, which by their nature have to build affordably, have shown time and again that the key to good design is simplicity. A well-proportioned building needs no gob-ons, but an ill-proportioned one will stay ill-proportioned no matter how many folderols you paste on to it. Good materials *can* cost more but, given the support of a determined planning authority, higher prices are not inevitable. The developer simply compensates for the increased building cost by paying less for the land.

Usually where this has happened the land price has dropped by around 10 per cent. Given that planning permission may increase the value of farmland from £3,000 to £250,000 an acre, and that raising the building specification will trim the margin from 8,200 per cent to just over 7,400 per cent, the vendor can still weep all the way to the Bahamas. Unfortunately, local authorities following this approach are in a small minority. The majority continue to demonstrate the inherent weakness of discretionary powers, which are only as good as the councils that wield them. Where powers are not invoked, the landscape is no better off than it was under Circular 22/80. This is not what we voted for. The Labour Party's pre-election promise, before it came to power in 1997, was to put 'good architecture and design at the centre of all government policy'. Its environmental policy document, 'In Trust for Tomorrow', looked forward to a future of health and cleanliness in which, our chests puffed up with meadow-sweet air, we all pedalled our bicycles to the railway station. The planning system would be strengthened and democratized to end the bias in favour of developers. There would be a clean-up initiative to recover derelict and contaminated land. An environmental Ombudsman would be appointed. There would be a special environmental division of the High Court; environmental studies would be essential to the school curriculum; there would be a legally enforceable right to roam in open country, moor and mountain. If not quite a Nimby's charter, then there would be at least a Nimby's safety net in the form of a third-party right of appeal against planning decisions made in defiance of a planning authority's own local plan.

Fat chance. Take away its obsession with foxhunting and Labour's passion for rural protection roughly equates to the Conservatives' enthusiasm for lesbian car maintenance co-operatives. It's not on the agenda. Changes to the planning system have all been geared towards more centralized decision-making, continually chipping away at local people's power over their own lives. It is not the people of the south Midlands who came up with the 'sub-regional spatial strategy' that would place 370,000 new homes in the countryside around Aylesbury, Bedford, Corby, Kettering, Luton, Milton Keynes, Northampton and Wellingborough. Not the people of Essex, Hertfordshire and Cambridgeshire who asked for 500,000 more households in the M11 corridor, or the people of Kent who want 31,000 new homes added to the existing 40,000 in Ashford. Not the people of the South-East as a whole who determined that their already overcrowded region should be the focus of the national housebuilding programme for the next twenty-five years.

Arguments about housing are complex, heartfelt and bitter. In 2003 the Treasury and the Office of the Deputy Prime Minister commissioned a report from a leading economist, Kate Barker, on the perceived problems of availability and price. Her conclusion was that, for the benefit of the wider economy, house-price fluctuations should be damped down and more should be done to provide affordable homes for people on lower incomes. To achieve price stability by increasing supply, she argued, would mean more than doubling the current rate of housebuilding – a prospect that the CPRE, among others, contemplates with no less horror than Betjeman and Williams-Ellis evinced towards the encircling tentacles of the urban octopus before the Second World War. Depressingly, the problem in the affluent early 2000s sinks its roots into the same barren soil as the post-slump 1930s. The physical decline of many towns, and particularly the degradation of the former industrial and mining centres of the North, have created a social and economic vortex that sucks people southward. In the South itself, it sucks them out of the towns and into the countryside. Thus we have created a vicious cycle. By fleeing 'rubbish' urban environments, better-off people ensure an accelerating spiral of decline in which the less attractive areas go into freefall. The only people willing to live in them are those who have no choice, and who lack the means to improve or maintain them. As homes fall vacant they are siphoned up at less than rock-bottom prices by landlords who install tenants on housing benefit, with dereliction and boarding-up often not far behind. The result – especially in the North, North-West and parts of the Midlands – is the phenomenon of 'low demand areas', economic black holes where, it is alleged, no one wants to live. Structurally sound housing is then fed to the bulldozers, and traditional communities broken up (see p. 84 for an account of how this goes wrong in practice).

It is not obvious to everyone that the best answer is to pack more and more houses into the South-East. Its capacity is not infinite. If the pressure is kept up, then overcrowding will erode quality of life and the South-East, too, will fall into decline, thus increasing still further the appetite of affluent urban émigrés for greenfield avenues and crescents. And this is the heart of the matter. The architecturally illiterate, sod-the-neighbours sprawl that throttles so much of rural England has done nothing to ease the problem of affordability. The Thatcherite free-for-all in the 1980s, when market forces were unleashed like greyhounds against the rural hare, actually caused a slow-down in the national housebuilding rate. While builders hit the fields with low-density, executive-style breeding boxes, the real need – for high-density, affordable housing on brownfield sites in towns – went unanswered, and the overall building rate was halved. As CPRE points out, this has nothing to do with planning but everything to do with the collapse in public funding for housing since the 1970s. The result has been accelerated urban decline, a quickening of the exodus to the countryside, and increasing polarization between the urban poor and the suburban/rural rich.

Releasing more greenfield sites is not the answer. The concept of 'need', too, begs for refinement. Ninety per cent of the demand for owner-occupied houses – and almost all demand from first-time buyers – is met through the turnover, or 'churn', of existing houses, not new-build. In any case, the shortage of building land and the iron fortress of Nimbyism must count alongside cold-water fusion and 'open government' as one of the myths of the age. In September 2003, CPRE calculated that Britain's fifteen leading housebuilders between them already held a landbank with planning permission for 278,866 houses – enough to build a continuous terrace from Land's End to John O'Groats. From this it concluded, reasonably enough, that the real needs lay elsewhere: tax incentives favouring urban renewal over suburban sprawl; more funding for affordable or 'social' housing; a coherent national policy to reduce the yawning prosperity gap between the regions.

The alternative is more and worse waste. Waste of depopulated town centres and inner cities; waste of countryside; waste of economic potential in Wales and the English regions; waste of the South-East. The impact of new housing is felt far beyond the land it is built on. The average new home gobbles between 50 and 60 tonnes of aggregates. If the government's projection for 2016 of 4.4m new households is correct, and if each of these households translates into a new house, then it will take an alp's worth of aggregate – between 200m and 250m tonnes – to build them. To get this much out of the ground, says CPRE, some 75–90 new quarries would have to be gouged from the landscape. And that's before you even start thinking about the roads.

Building badly is not the only problem. As Betjeman and Williams-Ellis foresaw, and as decades of experience have confirmed, what we choose to demolish is every bit as important as what we build in its place. In July 2004, at the suggestion of SAVE Britain's Heritage, I visited the town of Nelson, in east Lancashire. It hugs the slopes of a winding valley beneath Pendle Hill, the still-lovely eminence upon which, in 1652, George Fox is supposed to have had the vision that inspired Quakerism. What fired him then was hypocrisy – the discrepancy between what Christians professed to believe and the way they behaved in practice. Not much changes. It was not in the name of God but in that of John Prescott that the valley now lay in the way of a policy juggernaut spilling fine words but fixed on destruction.

The fine words were about 'sustainable communities' – a phrase that even its authors were hard-pressed to define, though it seemed to involve house-price inflation and access to jobs, schools, shops and 'leisure facilities'. Where such things were lacking, the theory went, the community needed to be 'regenerated' or 'renewed' – words that, in the minds of folk not schooled in weasel-speak, were often taken to mean restored or repaired. The reality could not have been more different. To sustain such a community, said the Office of the Deputy Prime Minister (ODPM), you first had to knock its houses down and disperse the people. Not only here in Nelson, but in towns and cities right across the Midlands and the North. Houses by the thousand – 1,700 in Salford, 2,000 in Newcastle and Gateshead, 2,700 on Merseyside, 1,600 in south Yorkshire, 790 here in east Lancashire – were being lined up for the bulldozers in an agonizing, slow-motion replay of the community-wrecking slum clearances of the 1960s. Between 1.5m and 2m houses in nine separate areas were caught in the crusaders' sights as targets for the un-improvably Orwellian process of 'housing market restructuring'.

Not far from Nelson, in Darwen, a builder who had just refurbished his house from roof to kitchen opened a letter from the borough council, signed by the 'Director of Regeneration', and found it had been declared unfit for habitation. Further east, in the Toxteth area of Liverpool, the same questions were repeated over and again. How did you sustain a community by knocking its houses down? Where were people supposed to go? In the minds of residents there was duplicity – local authorities sheltering behind legal language in order to justify the condemnation of sound and healthy houses. There was confusion. Officials in Darwen, for example, were telling owners of condemned properties that they could not discharge unfitness notices by carrying out the repairs specified in them (if they were 'unfit' on the day of inspection, then that was the end of it). ODPM, meanwhile, was telling me exactly the opposite. Everywhere, people felt they had been misled by cunningly worded letters that didn't mean what they seemed to say. In Nelson they were ready even to believe their local authority guilty of harassment – of orchestrating acts of vandalism and assault in order to drive them out of their homes. It showed the depths they had reached.

Planning offensives, like military campaigns or space probes, are given virile, go-get-'em titles. Thus it is that the nine areas targeted for 'regeneration' share with NASA's Mars mission the title of 'Pathfinder', and that individual Pathfinders personalize themselves with names of their own. In east Lancashire the bulldozers are directed by 'Elevate' – a partnership of five local authorities in alliance with Lancashire County Council and ODPM. Twice already a public inquiry had rejected its scorched-earth proposals for Nelson's Whitefield ward, yet like hyenas driven from a carcass the men with clipboards still circled and waited.

Nelson, it has to be said, is not typical. It grew in the nineteenth century as a purpose-built, pre-planned mill town, linked to the superhighway of Victorian industrial enterprise by the Leeds and Liverpool canal. Its mills and cottages were beautifully built of Lancashire sandstone; its streets cobbled in granite; its pavements clad in York stone flags. Every block of stone was flawlessly cut, immaculately laid and mortared; every terrace subtly different from its neighbours. Like all the best old towns, it folds into its landscape, the terraces like natural contours girdling the hill. Its exceptional 'heritage' quality marked it out and attracted some high-profile defenders. English Heritage, the Ancient Monuments Society, the Victorian Society, the Council for British Archaeology, the Heritage Trust for the North West, SAVE Britain's Heritage and the Prince's Foundation all lined up behind the campaign to oppose demolition. In October 2003, the Prince of Wales himself arrived to offer his encouragement.

But still the threat would not go away. The grass beginning to push up between the cobbles in Portland Street might have been thought attractive, but not so the nailed-up doors and windows. The ugly, corrugated metal shutters bore the stencilled confession of their perpetrators – 'Pendle Borough Council'. This, and other streets like it, were the muted victims of an 'urban renewal' scheme introduced by the council four years earlier. By 'renewal', in the language of the day, was meant 'clearance'. A line was drawn around the houses to be demolished, wiping out a third of the old mill town. Pendle Borough Council issued a Compulsory Purchase Order and sent out letters inviting the owners of 146 properties to sell up to them. Many agreed. The Asian families who arrived in Nelson in the 1960s had a traditional respect for authority, and felt it their duty to obey. Others were encouraged to sell by estate agents who advised them to make haste. The longer they stayed with properties emptying around them, the deeper would become the blight and the lower the 'market value' by which the compensation would be set. What would be the value of a lone survivor, pointlessly holding out in a blighted, boarded-up street? Heroism always carries a cost. That's what makes it heroic.

Repair work ceased in the condemned streets; roadways sank into their drains; backyard walls were pulled down; York stone flags were prised up and carted away. Blight spread like a fungus. Even so, the community had enough strength left in it to fight back. In 2002 came a public inquiry, which might be characterized as Pendle Borough Council versus the people. The people won. They loved their houses, they said, and they loved their community. The inspector agreed, and recommended to the deputy prime minister that the compulsory purchase order should not be allowed. The houses should be repaired, not cleared.

By now, however, the government had come up with its Pathfinder scheme. Across the Midlands and North as a whole, it intended to spend hundreds of millions of pounds on 'housing market renewal' – £150m in the first financial year, rising to £450m in 2007–8, and ever upward for ten or fifteen years – with nearly half the money in the first two years earmarked for clearance. The people of Nelson unwittingly found themselves in the way of a massive programme on which political reputations might hang, and on which enormous sums depended. Central to Pathfinder policy was the theory of 'market failure'. This was said to have occurred in places where property prices were low, where there were lots of empty houses, where there was a rapid turnover of owners or tenants and a high number of homes for sale or let. The root of the problem was an over-supply of Victorian terrace houses that nobody wanted to live in. The market would not recover until they were significantly reduced in number and replaced by something a bit more twenty-first-century. Hence the notion of 'housing market renewal' and its apparently perverse emphasis on demolition. Hence, too, the borough council's inelegant stampede for the bulldozer. Like all the local authorities in the economically stricken Pathfinder areas, it desperately needed money. Like the others, too, it believed the ODPM windfall, driven as it was by conviction of market failure, depended on a demolition programme.

There was only one way it could go. Having been de-populated and boarded up, Whitefield was full of empty houses – 'voids' in the awful jargon of government. A few properties did change hands, often to speculators who were happy to keep them empty Increasingly, voices in the street raised only their own echo. The market meanwhile was demonstrating a mind of its own, with price ripples spreading steadily from south to north. According to the market research organization Hometrack, by April 2004 Lancashire was experiencing some of the highest price increases in the country. Local newspapers reported rises of 50 per cent over two years in Burnley, and 44 per cent in Pendle. The sums offered to people to move out – often less than £10,000 per house – seemed worse than inadequate. Where could you buy another house for that?

John Prescott did not actually reject his own inspector's recommendation but, unprecedentedly, asked him to reopen the public inquiry and think again. He wanted to know whether the inspector's conclusions held good in the face of market failure. Again the protagonists hurled themselves into the fray. Again the inspector found for the people. The council dredged up every argument it could think of – the homes were unsuitable for modern living; they were beyond repair; nobody wanted to live in them; their 'heritage value' was negligible – but it was a farce. Under the forensic probing of lawyers for the residents, English Heritage and the rest, the council's case fell apart. The letter from John Prescott that hit the desk of its senior solicitor in September 2003 spelled it out in humiliating detail. The council had provided no convincing evidence of houses left empty over long periods before its own policies blighted the area; no evidence of high population turnover; no evidence that clearance would make better sense than re-habilitation, or that the council had understood the historical importance of what it wanted to destroy. And there came a critical thrust deep into the heart of Pathfinder policy itself. Could you 'sustain' a community by dispersing it? Prescott seemed to have no doubt:

> It is likely that many families would be forced into accom-
> modation elsewhere as they would be unable to afford the
> replacement housing that would be provided. [The Secretary of
> State] agrees that retention and renovation of the present
> dwellings would be more likely to promote continuing
> community cohesion.

In a speech two months later, he said it again: 'I believe passionately in the value of our heritage and the need to preserve old buildings. In the past, regeneration has often meant wholesale demolition. But demolition is not an essential part of regeneration.' He praised a scheme in Salford, where Urban Splash was 'turning the inside of old terraced houses upside down to create attractive modern living spaces in a traditional Victorian house'; and, in case anyone still had not grasped it, he hammered the message home: 'That's also why I refused permission for the wholesale demolition of Victorian terraced housing in Nelson in Lancashire.'

But still Pendle Borough Council would not lie down. It got itself a new Executive Director (Regeneration) and a new tactical plan. With the aid of consultants, it would produce an 'Area Development Framework' that would pay full attention to the people of Nelson and give them what they asked for. To see what this meant, I visited a neighbourhood drop-in where Whitefield residents were invited to inspect the consultants' proposals and record their preferences. According to the council, it was all the result of lengthy public consultations in which local people were 'physically, positively involved'.

What did it actually amount to? Seven residents of Whitefield and two councillors had accompanied consultants on a 'walkabout'. There had been five 'stakeholder interviews' with representatives of local organizations; a 'workshop' attended by 27 residents, 12 agencies and two councillors, and 'outreach sessions' involving 166 people in all. The consensus, reported by Social Regeneration Consultants (SRC), was for a policy of 'patch and mend' – i.e., in SRC's own words, 'for the area to be improved through refurbishment and a limited amount of selective demolition but not through wholesale clearance and redevelopment'. The result?

Demolition.

Armed with SRC's research, the principal contractor, Nathaniel Lichfield and Partners, had come up with four options. The least radical of these involved knocking down 95 houses; the most invasive called for 385. Maps were blocked in appropriate colours, and lists of addresses were printed of those thought likely to be in the firing line. The ODPM later would not tell me whether or not this constituted the 'wholesale demolition' John Prescott claimed to have banned, but the dislocation between the popular preference for 'patch and mend', and Pendle Borough Council's harvest of rubble, was evident for all to see.

Meanwhile, similar things were happening nearby in Darwen. Lee Woods, a tattooist with a passion for boxing memorabilia, had restored a pair of adjoining houses in Redearth Road to a standard that Adam Wilkinson, Secretary of SAVE Britain's Heritage, described as 'pretty impressively nice'. It had a gym and a sitting room with trompe l'oeil stone walls and a log-burning fire, designed with unconscious irony to suggest the interior of an Englishman's castle. Not far away in Lower Cross Street, Simon Gilmartin, a builder, had poured £18,000 and a great deal of professional skill into a house that he let to a tenant. It had new windows, new damp course, new plastering, wiring and plumbing, new central heating, new bathroom. It had been reroofed, and the outside walls had been sandblasted and repointed. Only a year earlier he had been given a twenty-five-year mortgage by a building society whose surveyor described the house as a 'recently refurbished example' in a 'popular residential area'. On the basis of ten-minute visits from a young surveyor with a clipboard, both men's houses, along with those of 147 of their neighbours, had been classified as unfit for habitation by reason of 'serious disrepair' and scheduled for clearance.

Adam Wilkinson was one of a group from national and local heritage bodies that walked around the area and inspected the houses. 'The whole exercise,' he said, 'is a complete farce' – a judgement reinforced when a group of residents arranged for their houses to be examined by a past president of the Institution of Structural Engineers who – unlike the council's men who had not entered roof spaces or looked beneath floors – conducted full structural surveys. His opinion was unequivocal. In the worst case, a property with a likely market value of £60,000 needed perhaps £4,000-worth of modernization work. Four others needed maybe £2,000 or £3,000 spending on them; three needed nothing. All were sound, with at least thirty years of healthy life ahead of them; all had been reroofed; none was unfit for habitation; there was, in short, nothing seriously wrong with them. Professional verdict: 'Although regenerating the area may be an honourable objective, it is dishonourable to suggest that these houses are not habitable.'

It may also be illegal. Twice I asked ODPM, and twice I got the same answer. No property that is not unfit for human habitation may be included in a clearance area. No property may be classified as unfit unless it has been the subject of a full structural survey. Despite this, and despite members of Blackburn with Darwen Borough Council 'calling in' the plan for reappraisal, there was to be no retreat. The residents began to form themselves into committees and called in advice from Nelson, but they felt like a coracle against the ship of state. The bulldozers would roll in, and the community thus 'sustained' would be replaced by shops, offices and a school.

With its high-profile friends and interest from the national press, Nelson's much fought-over Whitefield ward enjoyed an upswing in its fortunes. Pendle Borough Council announced that it had repented of error, had seen the light and was – as one councillor put it – 'under new management'. In November 2004 an 'Inquiry by Design', conducted by the Prince's Foundation for the Built Environment, produced a Whitefield Masterplan that, for the first time, had a chance of pleasing governors and governed alike. It proposed a small amount of demolition – something in the region of fifty houses looked likely – with a much stronger emphasis on conservation and constructive renewal. Some of the old houses would be knocked together to make larger units; a magnificent canalside textile mill would be refurbished and reopened as workshops for high-tech and light industrial use; new apartments and a new pub/café would be built beside the canal; there would be a marina and new square; blocked-off roads would reopen so that the ward would reconnect with the town; the redundant parish church with its commanding spire – the second highest in Lancashire – would become the headquarters of the Heritage Trust for the North West. Most importantly, the lovely stone terraces, beautiful as any in England, would be sensitively refurbished, not flattened. The community would not only be sustained but enhanced. Whitefield, in all its richness and diversity, could fill its lungs and breathe again.

Well, let's hope. An awful lot of investment has to be found – the initial cost estimate was £100m spread over twenty years – and the Masterplan was only an idea for the council to consider, not an adopted policy. But everyone's talking a good future and it may even happen that some kind of precedent has been set for other places. When I told the story of Nelson in the *Sunday Times Magazine* (the foregoing paragraphs being adapted from what I wrote), there were a number of immediate results. Pendle Borough Council and the Elevate Pathfinder wrote letters of complaint; so did the Minister for Regeneration, Lord Jeff Rooker. No surprise there (though it was a pleasure to be able to quote the deputy prime minister in defence of what I had said). But I also heard from people in other communities across the Midlands and the North, who were similarly threatened with the sustaining intervention of a Pathfinder wrecking ball. They were confused, frightened, hardly able to believe that what was happening to them could be true, desperate to know how they could defend themselves. I put them in touch with each other, hoping that, in unity, they could find some kind of strength. But they were ordinary people in ordinary homes. Their houses did not have the 'heritage' cachet of Nelson's, and so lacked the support of high-profile, royally connected defenders. One cannot feel optimistic for them. Neither should one forget that, even at

Nelson, and despite the council's avowed appointment with the blinding light, if the planners had got their way, the bulldozers in Whitefield would already have done their work and the community would have been dispersed.

This is the tragedy at the heart of Pathfinder policy. Investment in run-down urban areas; new administrative networks; consultation with local people about their needs; demolition of unsafe or unwanted buildings. Who could argue with any of that? The paths ought to have been strewn with palms. That they were strewn instead with angry protests was an indictment of a policy that placed ideology ahead of people and put too much weight on a single article of faith – that wholesale demolition of nineteenth-century housing was some kind of twenty-first-century panacea. And all this – the quota-driven waste of people's homes, the pretence that communities were somehow served by being broken up – was going on at exactly the same time as Mr Prescott's department was forcing half a million new homes on equally uncomprehending and devastated communities along the so-called Eastern Corridor around the M11 and Stansted Airport. Both exercises are conducted in the belief that it is both possible and proper for the government to influence the housing market simply by manipulating the supply rate. Prices in Lancashire are too low. Answer: knock a few down and increase the demand for the remainder. Prices in the South-East are too high. Answer: build a load more and make 'em cheap. Any local people who object can be dismissed as Nimbys.

There is no point now weeping about transport policy. The 'social railway' died with the Beeching cuts of the 1960s, and Margaret Thatcher's 'great car economy' in the 1980s only accelerated the pace of what was already happening. Motor traffic is doing to the countryside what it has already done to the towns, turning roads that once united communities into barriers that divide them. Many villages now have rush-hours. Unluckier ones have heavy traffic round the clock. By the most pessimistic estimates, the volume of traffic in some places will have trebled by 2025. Congestion and danger are not the only threats. The spoiling of villages and lanes, like the disfigurement of beautiful buildings, has a corrosive effect on both public and official attitudes. We fall into the habit of acceptance. What's another roundabout when there are already so many? Even the landscape has become disposable: improved 'access' means improved likelihood of development permission.

Huge sums (over 12 per cent of the total local transport budget in 1996/7) were spent on widening roads and strengthening (i.e., disfiguring) old stone and brick bridges in preparation for the arrival of forty-tonne intercontinental lorries. As traffic weights, speed and volume increase, so panic takes hold. How is the monster to be appeased? Once-tranquil villages get high, suburban-style concrete kerbstones, road humps, chicanes, road narrowings, yellow lines, bollards, pedestrian refuges, street lights, traffic lights and car parks – very often on what used to be the green or square. Roundabouts in particular are like Trojan horses from Basildon. Out of them spring huge lighting columns; then come the sign erectors (roundabouts attract them like nothing else) and the filling stations, ultimate junk architecture, designed with the visual equivalent of a loudhailer. Oil companies don't want their premises to blend in; they want them to stand out. It is corporate identity, not local distinctiveness that counts, and it is expressed in a language calculated to reach across continents. Close behind them come fast-food joints whose style has less to do with the shire counties of England than with the freeways of Middle America.

But road trash is not exclusive to privateers. Look at what highways departments do. The job of highways officials is to make sure that drivers know which way to turn, where to slow down or stop, where they may or may not park, and what they should beware of. The usual technique is overkill. If a village has a speed limit, then the 'gateway' sign will be followed by smaller repeater signs at 250-metre intervals until the restriction ends. If the main-street limit is 40mph, then the side roads will usually be reduced to 30, with each junction flanked by yet more signs. The result is that you often get more traffic signs in a village high street than you do in a town.

If the place is on a tourist route, then the problem will be exacerbated by a blizzard of huge, reflective directional signs which borrow their style and character from trunk roads. Even the remotest one-horse junctions have had their traditional fingerposts uprooted (new ones were banned between 1964 and 1994) and replaced with gigantic metal horrors which, with all the apparent urgency of a major urban intersection, often have nothing better to say than 'Byroad'. It is one of the identifying marks of a system in chaos – of inertia masquerading as action. You see it in ill-run organizations of every kind – schools, shops, offices, factories, public places: when all else fails, stick up a notice. This way, that way, go faster, go slower, don't stop, keep out, no photography, no turning, no dumping, no feeding the animals, park and ride, toilets. To its immense credit, the CPRE has long campaigned against this tideline of rubbish, and has even managed to persuade some local authorities to put back the fingerposts.

Add to all this the pub and shop signs, pylons and tele-communications masts, and you have a rural landscape that is as much like its original self as a Pot Noodle is to an Aberdeen Angus. It doesn't have to be like this. The bottle bank does not have to be the most prominent feature in a village centre. Rural signposts do not have to be readable at 70mph. Sense and sensitivity, rather than stopwatch and traffic counts, can reduce the number of changing speed limits, with all the clutter and confusion they cause. With a single, uniform limit throughout the village, you can do as the French do and sign it at entry and exit points only. Telecommunications companies are in-eradicable features of early twenty-first-century life, but they don't all need their own dedicated masts; they can share with each other, or even make use of existing structures. Everywhere is *somewhere* – towns and villages are not just markers on a route. Traffic can be accommodated and controlled without the ritual sacrifice of character and life. It is now widely acknowledged that 'soft' traffic controls often work better than hard architecture. Narrow roads with bends in them, for example, can do more to improve driver behaviour than wide straight ones – an idea used to good effect in the Prince of Wales's 'model' village at Poundbury in Dorset. You do not have to like the derided (by architects) 'pastiche' vernacular architecture to recognize and admire the planning here, though it remains beyond the grasp of many local high-ways departments in which straight-line geometry has the addictive power of laudanum.

Here and there are little flickers of hope. Even at Milton Keynes quite recently I saw plans for a new 'home zone' that showed narrow streets with curves in them to make speeding impossible, houses arranged in discrete 'farmstead-type' groupings rather than lined up like spectators at their own funeral, and parked cars kept out of sight behind them. The nasty 'pinch points' and absence of vistas lacked some of the subtlety and craft of the original, but – although the planners would rather have their toenails plucked out than admit it – in every meaningful respect the thinking was pure Poundbury. Elsewhere, some of the best new ideas have come out of historic town centres where, no less than in the countryside, jarring notes create disharmony. If the eye is distracted, it misses as much as it sees. Bury St Edmunds, in Suffolk, offers a good example of modern thinking in a traditional package. In the historic centre there are no concrete lamp standards. Cobbles and York stone paving have been introduced in Crown Street and Angel Hill; granite setts instead of white lines have been used to mark parking bays; crushed stone in epoxy resin has been used to create a hardwearing but handsome 'carriage-drive' surface in Chequer Square; Hatter and Whiting Streets have been narrowed to slow the traffic and make more room for pedestrians. Along with Shrewsbury, Lincoln and Halifax, Bury pioneered less intrusive ways of accommodating, directing and controlling town-centre traffic. Out went the horrible yellow lines, speed bumps, chicanes and oversized illuminated road signs. In came 20mph speed limits, weight restrictions, pedestrian priority zones, smaller, fewer and better-designed road signs. 'Gateways' at each of the Core Zone's main entry points signalled to drivers that they were entering an area where different rules applied. Similar ideas have been scaled down for use in some villages – 'gateways' and different, gravel-look road surfaces to indicate a change of environment, and an absence of white lines to distinguish village streets from the 'open road' (despite the incredulity of some highways departments, traffic slows down rather than accelerates when the central line is removed). Progress is slow; standards remain poor – even quite minor rural junctions still attract a dozen official road signs – but let no one say they haven't been offered a better example.

What no one has been able to put back, though many have tried, is the red squirrel. If the trashing of the country-side has weighed heavily on the human spirit, it is nothing to what it has done to other life forms. The litany of loss – of plants, insects, fish, reptiles, amphibians, birds and mammals driven either to the edge of extinction or forced into small pro-tected enclaves – is a tragic catalogue of a war pursued against an axis of innocence. In ignorance or arrogance (in this context, much the same thing), or in blind assumption of our God-given right of dominion, we have rubbished nature itself. Our power, temporary though it may be, is purely destructive. We know how to end life. We know a great deal less about how to restore it.

The red squirrel is a prime example. Unlike many less obviously appealing creatures it has never lacked for human friends, or for well-meaning efforts to save it. To see what goes wrong you need look no further than Regent's Park in central London. It was here in the early 1980s that the Zoological Society of London decided to re-establish a colony of reds which, as in every other corner of England, had long ago been driven out by the larger and more aggressive greys. The plan looked OK. What had killed the reds was their failure in competition for food. To stop it happening again, all their protectors had to do was make sure that they had plenty to eat. But alas . . .

Unlike the grey, the red squirrel does not like crossing from tree to tree along the floor: it prefers what Regent's Park, not being a self-regenerating wild woodland, specifically lacks – continuous tree canopy. Worse, its last experience of Regent's Park had been in the age of the carriage horse. You might as well have dug up Dr Johnson and expected him to drive to Gatwick Airport around the M25. It all ended, with a *thunk* and a *miaow*, in tears. Within weeks most of the squirrels had been either run over or eaten by cats, and the few survivors were rounded up and sent to a zoo in North Wales.

Other attempts have been better thought through, but no better favoured. Giving squirrels a new home is one thing; making them thrive is quite another. The problem is not lack of any suitable environment but the animal's inability to withstand stress in transit, and its poor defence against viral diseases such as parapox. For a while, hope rested on a well-publicized colony in Thetford Forest Park, Norfolk, but this too was hit by disease and the project was suspended (though there were three confirmed sightings of red squirrels here in 2004 – one road casualty and two happily alive).

This is the essence of the Rio Earth Summit. If a species is to re-establish itself following local extinction, then the whole ecosystem must be conserved and the causes of extinction eradicated. Sadly, in the case of any widespread reintroduction of the red squirrel, this would mean killing all the greys – something that would be practically impossible if not politically unthinkable. The problem with the grey is that it is

101

the mammalian equivalent of a weed – a species that occupies a safe and useful niche in the ecology of its natural range, but becomes a rapacious and destabilizing nuisance when transplanted. A classic example is the invasion of West Country hedgerows by rhododendrons – a sci-fi nightmare of unarmed natives devastated by aliens. Another was the wrecking of riverbanks by coypu before they were trapped to extinction in the 1970s. The same waters are now being invaded by mink, typically vicious aquatic cousins of the weasel, that were 'liberated' from fur farms by animal rights activists. They saved the mink all right, but only at the cost of condemning the water vole (Ratty of *The Wind in the Willows*) to the danger list. Nature once affronted is hard to reconcile, and it will make no distinction between good intentions and bad. In 1974, for example, the apparently benign and eternally popular hedgehog was deliberately introduced to the island of South Uist. This was heaven for the hedgehogs, which rapidly multiplied and spread to other islands, but hell for the ground-nesting birds whose eggs the incomers feasted upon and whose numbers declined as rapidly as the hedgehogs' orgiastically soared. Conservation it was not.

You see the same thing the world over – the wrong organism in the wrong place. It is a shock to British sensibilities to find that 'pests' in Western Australia include blackbird, bullfinch, turtle dove, skylark and song thrush, and that their fate is to be treated as vermin, worse than rubbish. In New Zealand, too, ground-nesting birds have been devastated by weasels and stoats imported by Victorian immigrants. 'Nature' is a hard thing to define, not least because we tend to exclude ourselves from it. A termite hill is 'natural'; a tower block is not. Predation by weasels is God-given and right; hunting by humans is wrong. In arguments about GM, embryo research, cloning and other issues where human ingenuity gives fresh spin to old forces, we struggle to decide what is 'natural' and what is not. It is an argument that spills out on to the fields. The landscape we defend is not natural wilderness. It is our own creation, and the haven it once offered to wildlife was a by-product rather than the intended outcome of earlier farming methods. How, then, can we say what is right and wrong? Now that we have little more use for a stockproof hedge than we have for a traction engine, the need is to find a new equilibrium which, for the sake of our sanity, balances sentiment with utility and measures the value of landscape in something more than just crop yields. We owe more to ourselves than to lay waste to our own finest creation.

For centuries in lowland England the point of balance, the perfect interface between man and nature, has been the hedge. England's hedgerows are like its bread, beer and cheese – basic, everyday staples one should be able to take for granted but which, thanks to the homogenizing effects of mass production, too often have to be fought for. What is the first thing any Englishman will do when he has a plot to call his own? Plant a hedge. Yet more than 200,000 miles of them have been ripped out since the war – enough to girdle the earth twenty-five times, and almost exactly equal to the length of hedge planted during the Parliamentary Enclosures between 1750 and 1850. The earliest hedges were fillets of primeval woodland left by Neolithic settlers as they carved the ancient forest into farms. No one knows quite when the point was reached where nature could no longer meet the demand and men had to start replacing what they had removed, but it seems likely that hedge-planting was quite commonplace by the tenth century.

A written record of 940 notes 'the hedge that Aelfric made' at Kington Langley in Wiltshire.

A Saxon farmer reborn in 1939 would have blanched at the encroachment of towns. Away from the smoke and the suburbs, however, he would have recognized his own footprint, still clear and sharp in a landscape of small, eccentrically shaped fields parcelled by hedges. A Saxon farmer reborn thirty years later would have wondered what planet he was on. The post-war drive for maximum food production, the replacement of ancient 'organic' land management techniques by chemical controls, the development of vast machinery that depended for its efficiency on huge fields of geometrical exactitude, had shaved away two-and-a-half millennia of landscape history. Small tenant farms were 'taken in hand' by the estates that owned them; ponds and marshes were drained and levelled; animals were crammed into sheds and yards to make way for the tide of cereals. Where once had lain a rich mix of English acres, there now stretched arable *hectares*, each one as blandly and bankably predictable as the last. Anyone who wanted to connect with the past would have to hire an aeroplane and take photographs of crop marks. Where old trees survived as relics of forgotten hedges or copses, they were marooned in the middle of the desert like dying camels, their metabolisms sent haywire by alternate drifts of fertilizer and herbicide.

Before the hedge, ever since the last retreat of the ice, Britain had been a land of trees, and nowhere do feelings of atavism swell up more powerfully than in one of the few surviving fragments of the ancient wildwood. An hour here is like an hour transported from a fairy tale. The flutter of unseen wings; the narrow slantings of light; the rustle and stir that may or may not be the work of the wind; the sensation of being observed, from every angle, by a multitude of eyes. It makes children of us all. Like the sea, the forest transcends human intellect and hands the advantage to creatures of harsher instinct. Like the sea's, its mysteries are felt rather than understood. The trees are where we came from. They led us to fire and the wheel. They are the root and kernel of our history, the scene of our earliest worship, givers of light and energy, the source of all that we love and fear.

In the less sentimental centuries that preceded our own, before 'landscape' became collectable, woods were things of utility, tools of the economy, not masterpieces of nature to be cherished for their loveliness. Until the First World War they were the nation's power stations, providing wood and charcoal for burning; and its builders' yards, supplying timber for houses and ships. Oak bark was used in tanning. Hornbeam made gear-wheels for millwrights; ash went into wheel spokes and tool handles; hazel made hurdles for shepherds, spars for thatchers and wattle for wattle-and-daub. The queue at the forest gate was never-ending: cartwrights, furniture-makers, boat-builders, carvers, clog-makers, mine-owners, millers, smiths. Woodland crafts and trades still float in the alphabet soup of the telephone directory: Ashburner, Carpenter, Cooper, Fletcher, Forrester, Hedges, Hurst, Lee, Joiner, Sawyer, Turner, Wheeler, Wheelwright, Wood, Woods, Woodward . . . A gazetteer of place names tells the same story. Any address with, for example, *leigh, ley, hurst, shaw, holt, weald, wold* or *wood* in its name denotes the site of a wood, copse or clearing. On a domestic scale, we are still at it. On a short drive near my home in Norfolk I pass The Larches, The Walnuts, The Beeches, The Maples and The Laurels. It survives, too, in our regular challenge to strangers: 'What brings you to this neck of the woods?'

But the modern world has stood the ancient culture on its head. To wield an axe now is an affront to public sensibility. Nothing unites us like an endangered tree. In the face of an advancing bypass, Druids link arms with executives' wives. The older the tree, the greater the offence. When some 150-year-old oaks were felled in Windsor Great Park in 1995, critics likened the outrage to 'chopping down the Queen Mother'. The irony is that most of the tree-loving public can't tell beech from birch and think hornbeam is a dance performed by sailors. Once we could read trees like city signposts; now we are as blind to their particularities as we are deaf to the distinguishing subtleties of birdsong. Holly, maybe, is one of the few that most of us recognize, though we forget why so many of them were planted (as ploughmen's markers), or why gypsy women liked to give birth beneath them (they gave good shelter), or that they were the preferred material for driving-whips. Who now remembers that fresh lime leaves make a good sandwich; or that maple makes the best harps; or that yew trees are often older than the medieval churches by which they grow; or that in the dark days before the Pill, when every romantic encounter was a trapdoor to disgrace, wild juniper was everywoman's answer to an unwanted pregnancy? In our blind love for unnameable trees, however, lies a solid plank of hope. More charities exist for the planting, care and maintenance of English trees than exist for the succour of English orphans. By restoring them to the landscape, or so we hope, we become nature's midwives, attending to the rebirth of what we have lost, restocking the empty ark. We can even do it in our gardens.

Trees in a sense are easy. You plant them; they grow. And, providing you have enough space and time, you can think as big as you like. The grandeur of oak never diminishes, even as you yourself grow older. With our common native fauna, on the other hand, there is an adjustment to be made. In the whole of nursery literature there is no villain more ruthless or chilling than the weasel; and none more disappointing. You grow up, swap *The Wind in the Willows* for a field guide, and find that this feared aristocrat of evil is hardly bigger than a chipolata. A small female may reach little more than an ounce and a half; even a big male will do well to outweigh a hamburger. Quake in your nest if you are a mouse or a baby songbird. Otherwise you've as much to fear from a battered haddock.

The reason we have demonized the likes of weasel, stoat and fox is not that they are truly frightening – even a fox in a fury would think twice about a tomcat – but rather that they are all that's left. Outside the human jungles of City or government, we don't have proper 'top predators'. Other countries have bears, wolves, tigers. We have minuscule furry morsels whose own fate, as often as not, is to end up in an owl pellet. This is why film crews spend more time on the plains of Africa than they do in Cowslip Wood. Why every fat black farm cat photographed on a fence-post brings out a posse of big game hunters crying panther. Why we lionize our heroes rather than weaselize them. Why we want wolves and lynxes back in Scotland. Well, dream on . . .

In the Stone Age we were outnumbered five to one by bears. For every human looking for food in the forest there were 800 wild boar, three wolves, six lynx. Sapient man might have been no match for wolf or cat in hunting deer, but when it

107

came to eliminating the competition he could make a hyena look philanthropic. Down came the forests; off came the meat and skins of their inhabitants. Bears clung on in Britain until some time between the eighth and tenth centuries; wild boar until the 1600s and wolf until 1743, when the last one was shot near Inverness after allegedly eating two children. It used to be thought that the last lynx was killed in approximately 300, but bones recently found in Scotland show that it survived there until at least the fourteenth century. For years there have been rumours of the wolf's imminent return to the Scottish Highlands or to the island of Rum. Zoologists argue that we have a moral obligation to reintroduce species that have been displaced by human interference. For them, this does not so much *include* the wolf as place it at the top of the agenda. Until we take steps to fully restore our own ecosystem, they say, we will lack the authority to tell other countries how to look after theirs. And of course you can't have a properly functioning, balanced ecosystem without top predators.

The imbalance in Scotland's ancient pinewoods is made obvious by the absence of young trees. Reason? Damage to saplings by grazing deer whose populations are out of control precisely because they have no predators. The Deer Commission for Scotland no longer conducts national head-counts, which it regards as unreliable, but its last estimate, for 2000–1, was 350,000 red deer and between 200,000 and 400,000 roe, plus significant but uncounted numbers of sika and some fallow. Despite the annual cull (76,000 red and 33,000 roe in that same year), numbers continue to rise and – because they are depleting their own habitat as well as damaging crops – they pass no rational test of sustainability. It's hard to see how this can go on. If the pinewoods themselves are not to be cast on to history's scrap heap, something will have to be done.

To the carnivore lobby, that 'something' is obvious – the reintroduction of the wolf. It argues that, outside Scandinavia and Russia, both of which are roamed by wolves, the 15,000 square miles of Scottish Highlands are the largest wilderness in Europe. Thus, it says, there is 'ample space for wolves to define territories and for dispersing individuals to meet and form new packs'. It points to a BBC phone-in that identified a 75 per cent public majority in favour of wolves. It is safe to suppose, however, that the 75 per cent did not include the Highland sheep farmers whose flocks to a wolf might seem like nothing so much as a lope-thru restaurant; and it was drily observed that most of the enthusiasm for wolves in Scotland came from well south of the border, and that no one ever suggests releasing them in the New Forest. Then, in November 2004, came news of an apparently more serious initiative, complete with willing landowner and financial backing. A multimillionaire businessman, Paul Lister, son of the co-founder of the MFI furniture chain, announced that, subject to public approval, he planned to introduce wolves to his 23,000-acre estate in Sutherland, forty miles north of Inverness. And not only wolves: through the same hills and valleys would roam brown bear, lynx, wild boar and European bison. In the shadow of the ancient Caledonian pine forest, he would recreate a wild fauna not seen in the UK for more than a thousand years.

But what would be the status in law of such animals? Could fencing them in an enclosure – even a huge one like the Alladale estate – be defined as a release into the wild? And could the enclosure contain them anyway? An 8ft fence needs only an 8ft snowdrift to become irrelevant. The political obstacles are of dizzying height, if not actually insurmountable. No wolf will get the freedom of the Highlands without the approval of Scottish Natural Heritage, which itself cannot act without the co-operation of landowners and the consent of the Scottish Executive, which has to issue a licence for release under the Wildlife and Countryside Act. This will happen when William McGonagall rises from the grave and is awarded the Nobel Prize for literature. 'I don't believe there is a community anywhere in Scotland that would agree to it,' said a government official. The Highland landscape for the most part is man-made by people who make their living from sheep, and what *Homo sapiens* has made, *Canis lupus* will not be allowed to put asunder.

The alternative might be to license the estate as a zoo, but the ethical issues are of a kind that would keep student debating societies arguing till breakfast. For the 'release' to be authentic, and for the restored wilderness to be true to its nature, the carnivores would need live prey. In the wild, this is tolerated by animal rights activists who accept that they cannot reform the God-given natures of beasts. But in a zoo? Feeding live deer to wolves? Can you imagine the clamour? (And this is before you even begin to consider the danger to humans of unscheduled encounters with wolf, bear or bison.) What it illustrates, if any further illustration were needed, is the extreme difficulty of re-establishing ecological integrity once it has been lost – like trying to find the point of balance on a toppled logan stone. Launching wolves against an overpopulation of deer is not like releasing predatory mites against pests in a greenhouse. Even if they could be persuaded to ignore the lamb and eat only venison, the sheer size of the deer population means they would have little impact. It perfectly demonstrates the kind of thing that happens when an ecosystem is shorn of a working part. One human intervention (killing wolves) can only be balanced by another (killing deer).

The argument will go on, possibly for decades. For all the talk of 'charismatic wildlife' and the free-spending tourists it might attract, the rebuilding of Britain's broken ecosystems has had to begin much lower down the scale. The dormouse may be diminutive in size but, much like the canary in the mineshaft, it is of huge significance. Dormice vote with their paws and do not linger in degraded environments. What they need is the kind of traditional coppiced woodland that more or less disappeared after the First World War. The result has been that, over the last 120 years or so, they have vanished from at least half their natural range. Even where they do hang on, their numbers are a fraction of what they were. Spreading naturally is not an option – they will not cross open country (even 100 metres of treeless land may be too much). Traditional woodlands now exist only in isolated fragments, islands in a sterilized sea of lifeless plough, and the lost hedgerows have cut the lines of communication. Their only hope, therefore, rests on yet more human tinkering.

So far there have been thirteen dormouse reintroductions – two each in Cambridgeshire and Yorkshire, one each in Cheshire, Suffolk, Warwickshire, Nottinghamshire, Bedfordshire, Buckinghamshire, Derbyshire, Staffordshire and Lincolnshire – of which all but one (Nottinghamshire) have succeeded. In Cheshire the original group of twenty-five to thirty thrived so well that the population has spread at least a kilometre from the release site. Well . . . that's all very encouraging, and nice for the dormice, but so what? They are tiny, reclusive and nocturnal, unlikely candidates for a sustained assault on biological waste. Even if the woods bulged with them you would be lucky ever to see one. And yet you would be in their debt – deeply so. Most of the 391 species covered by the UK Biodiversity Action Plan are in need of rescue for exactly the same reasons. Roadbuilding, urban sprawl and the despoliation of the farmed landscape have cost them their home, their food and their freedom of movement. At least 60 per cent of the BAP species depend on exactly the same kind of woodland as the dormice. Thus if you create an environment fit for a dormouse – which means, among other things, protecting new tree growth from the pestilential deer – you enrich the whole ecosystem. What's good for dormice is good also for lichens, mosses, fungi, flowering plants, butterflies and other insects, snails, mammals and birds. It is the ultimate special offer – buy one species, get hundreds more free.

And this is very much the point. Mammals and birds are the glamour end of the nature market – the window dressing that attracts the headlines. Reintroductions by the RSPB of ospreys, red kites and white-tailed sea eagles all deserved their front-page splashes. If the planned reintroduction of Scottish beavers succeeds, then they, too, will be camera magnets. Rather fewer column inches will be assigned to the pool frog, the black bog ant, the leaf-rolling weevil, the hazel pot beetle, the tadpole shrimp, the dotted bee fly, the nail fungus, the warty wax lichen, the northern prongwort or Appleyard's feather moss. Most of the BAP list is like this – species after species, all now rare, that were obscure even when they were common. Many of them have no popular names, only a broad description – 'a diving beetle' – and a taxonomic label,

Bidessus minutissimus. There are 25 species of ground beetle, 22 flies, nine ants, eight snails, seven wasps, six spiders, three worms, one sea squirt . . . Who needs them?

Well, who knows? Ecology is a young science with most of its work still left to do. Even at the current rate of loss, nature is so diverse and so complex that it remains literally unfathomable. It is not likely that we shall ever know exactly how each of the world's 10m (some say as many as 100m) species relates to all the others. The value of Rio lies in the recognition that nature itself is the best judge of what it needs, and that it is dangerous for us to impose an arbitrary scale of values based upon eye appeal. We must not cherry-pick; must not focus all our energies on the glamour species while ignoring, or shuddering at, all the rest. Still less should we imagine that we can loot without restraint – we cannot have our cod and eat it.

We learned the hard way not to use the sea as a rubbish dump. And we are learning the hard way that tipping radioactive waste, oil and sewage are not the only ways to trash an ocean. What you take out of the water matters just as much as what you put into it, and is just as degrading. Nowhere has man fiddled more frantically, and nowhere has he failed more catastrophically, than in his attempts to conserve fish. The history of the European fish fiasco – the piracy, the quota trade, the licensed overfishing – would fill a thick volume all on its own, and it would make happy reading only for lovers of tragi-comedy. Oceans, too, have an ecology. Big fish eat little fish, little fish eat the bigger fishes' babies, baby fish eat even smaller fish, and so on all the way down to plankton. If you break the chain, you risk losing the lot. The future is an empty sea, devoid of all life except the microscopic denizens of a fished-out marine junkyard.

As a conservation tool, fish quotas are about as much use as a panama hat in a monsoon – better than nothing, but only just. The problem for cod-crazy northern Europeans is that our favourite fish cannot be harvested cleanly, like barley or beans. It doesn't grow in isolation but congregates with others. If you trawl for cod, then you will also bring up haddock, whiting and flatfish. When you have exhausted your quota for cod but not for the others, you face one of those delightful European conundrums. If you go on fishing for whiting and haddock, as you are entitled to, you will also catch more cod. What, then, do you do with them?

You have a choice. You can land them illegally as 'black fish', risking the wrath of the law. Or you can dump them over the side as 'discards'. Either way the fish are dead, the breeding stock has taken a hit, and the quota system has failed in its purpose. By regulation we may limit how many fish are legally landed, but in mixed fisheries such as the North Sea it will have only an incidental bearing on how many actually die. Nor is this the only problem. 'Discards' include not only over-quota fish but undersized ones too. To preserve future breeding stock, the EU lays down minimum sizes below which fish may not be landed (35cm for cod, 30cm for haddock, 27cm for whiting). Unfortunately, baby cod don't realize they're not supposed to follow mummy into the net so they, too, die in unrecorded numbers. For a time, the European Commission favoured increasing the mesh size so that the young fish could slip through. But nature, alas, is not geared to legislative specifications. If the mesh were too large to hold immature cod, then it would also be too large to contain fully grown members of smaller species. You *could* conserve young cod, but only at the cost of never catching another whiting. This mattered because whiting is an economically important catch in its own right, and because (here's Catch-22) its favourite food is young cod. To let the whiting escape would guarantee not the survival of young cod but rather their slaughter by predators – like turning young wildebeest loose among hyenas.

The collapse of North Sea cod stocks, like the collapse of herring before it, was as predictable as the loss of pine martens from gamekeepered woodlands. If you kill an animal, and go on killing it before it has had time to breed, then sooner or later you'll find yourself casting your nets into an empty sea. In late 2002, the EU fisheries commissioner tried to limit North Sea fishermen to a maximum of seven days at sea per month. The UK fisheries minister, Elliot Morley, returned from Brussels after five days of hard argument to triumphantly announce that he had renegotiated us back up to fifteen days – a victory which, one can only hope, will turn out not to have been as hollow as it looked. In December 2004, in a last-ditch attempt to save the cod, the European Commission proposed that fishing boats should be banned from large areas of the North and Irish Seas. Almost at the same time, the Royal Commission on Environmental Pollution came to an identical conclusion. 'As a society,' it said, 'we have assigned much less priority to protecting the seas compared with the land. This needs to change urgently.' It called for radical change. 'Present policies have failed, and incremental improvements will not deliver what is needed. We face further collapses in fisheries and harm to the marine environment unless there is significant and urgent action.' Attitudes to the sea needed to be stood on their head. The age-old presumption in favour of fishing could not be allowed to continue: it should be permitted *only* where applicants for fishing rights were able to demonstrate and preserve the long-term sustainability of the stocks. In the immediate short term, this would mean a network of marine protected areas – approximately 30 per cent of UK waters – in which commercial fishing would be banned. The International Council for Exploration of the Sea (ICES) went even further, calling for total bans on cod fishing in the North Sea, Irish Sea, parts of the Atlantic west of Scotland, the Skagerrak, Kattegat and the eastern Baltic. It calculated that only around 40,000 tonnes of spawning cod remained in the North Sea, far short of the 70,000 tonnes usually accepted as the sustainable minimum, never mind the 150,000 tonnes recommended by ICES itself.

The EC, the royal commission and ICES were talking science. Ministers, however, have to talk politics. No one expected EU fisheries ministers to endorse total exclusion zones, and they didn't. The cost to their national fishing industries would be too high. Existing conservation measures were beginning to work. Closing the sea would be premature, and so on. So the fishing would continue, albeit against quotas and with fishermen's days at sea restricted, as before, to fifteen a month. Fishermen's leaders spoke of 'relief'. Marine scientists and conservation bodies saw it rather differently.

What makes the cod particularly precarious is its slow rate of maturation. If it dies before it's four, it doesn't spawn. If it doesn't spawn, then we are looking forward to a future of jellyfish and zooplankton. As it is, we are down to pollack, pout and sand eel. One of the theories that sustains hope is that if you exhaust one species you can always switch to another. Nowhere is this more evident than in the so-called 'industrial' fisheries that suck out vast tonnages of small fish to render into oil and fish-meal – fully 55 per cent of the weight of fish taken from the North Sea is destined for industrial processing rather than the frying pan. The ships involved in this ecological carpet-bombing are mostly British registered but all are Danish owned. Most of what they take is sand eel but, from the harvesters' point of view, it hardly matters. All they are interested in is protein. Huge nets bring to the surface vast quantities of tiny fish that are sucked up like silage and pumped directly into the hold. Each vessel is allowed a 15 per cent accidental 'bycatch' of larger species, though no reliable method exists for fisheries authorities to check the precise constituents of the resulting broth. What provokes anger is the sheer profligacy of a fleet whose principal end product is feed for salmon farms and Danish pigs. On average, it takes three tonnes of sand eels to make one tonne of salmon pellets – a deficit in the environmental balance sheet that dents the widely held theory that fish-farming is the ocean's fairy godmother. Already on the west coast of Scotland the Scandinavian-owned salmon-farming industry has convincingly demonstrated its supreme talent for environmental destruction.

Cross-infection of sea-lice from salmon cages has cut deep into the populations of sea trout and wild salmon, which in some rivers are feared already to be extinct. Norway's marine research institute calculates that 86 per cent of young wild salmon are eaten alive, or fatally infected with viral anaemia, by sea-lice. Wild fish suffered further damage in the 1990s when furunculosis, a virulent disease that causes internal bleeding and death by septicaemia or kidney damage, spread from Norway via Scottish fish farms. In 1998, an outbreak of infectious salmon anaemia resulted in the enforced slaughter of 4m farmed fish and the closure of 25 per cent of the Scottish industry.

Shellfish may be hit, too. A significant body of scientific opinion now believes that the increasing frequency of toxic 'algal blooms' around the west and north coasts is caused by over-enrichment of the water by pollutants from salmon farms. Widespread closure of mussel fisheries, necessary to protect human consumers from shellfish poisoning, is a regular feature of Scottish summers. But the worst pollutants are the farmed salmon themselves, which have escaped by their hundreds of thousands – a million in 1998–2003 alone. Not only do the truants compete for spawning sites but they debase the wild species by interbreeding. Farm fish may be hyper-efficient food converters but they have all the virility of lapdogs. They manage only a caponesque 16 per cent of the natives' breeding rate; and, lacking a home river, they have no instinct to return to it. Thus, instead of hurling themselves upstream, they hang about in sea lochs and estuaries with no very clear idea of what to do next. Norwegian scientists have worked out that the degree of genetic distinction between farmed and wild fish is being halved every 3.3 generations, with the unavoidable consequence that the 'wild' population eventually will be composed entirely of descendants from farmed fish. There will be fish in the water and fish on the table, but the wild North Atlantic salmon will be gone.

Whether you think any of this matters is largely a question of sentiment. We are a predominantly urban people.

Some are prepared to argue that we should be content with this: like battery hens, we have changed our natures. City streets, not fields and fens, are our environment now. Our journey into a technological future, insulated from the microbial crudities of unprotected contact with nature, is just another aspect of our Darwinian advancement into ever more perfect form. We don't need coppiced woodland, or water beetles, or the Adonis blue butterfly, or the medicinal leech, or peat bogs or kittiwakes or anything with a shell on its back, and what we don't need is, by definition, garbage. Exaggeration? Perhaps. But although it is not a thought that many of us will actually articulate, it is one to which most of us queue to pay service. Every time we go to the supermarket we wipe another voice from the dawn chorus. Between us, UK shoppers and taxpayers have paid to take more birds out of the sky than all the popgun battalions of pheasant-shooters and Mediterranean bunting-broilers put together. Since the mid 1970s, we've cut the blackbird population by 4.1m, yellowhammers by 2.6m, song thrushes by 1.9m, skylarks by 1.6m, linnets by 640,000, lapwings by 250,000 . . . a list so long and so familiar that it's becoming a bore. Everything we eat, or use, or throw away will have its impact. Every time we fail to recycle a bottle, or let a tap run, or buy farmed fish, we add another straw to the camel's back. Part of us accepts this, if only by choosing to ignore it. These things do not affect us personally; not directly; not yet. But another part of us feels a creeping, persistent unease. We struggle again with that same old question of naturalness – a struggle made insistent as nature itself responds to human disdain with its own forms of terrorism. Superbugs, bird flu, BSE, new variant CJD. This is why the biotechnology industry and the UK government face such powerful public opposition in their determination to impose GM crops. It's instinct, not intellect, that makes people watch a squirrel in a park. Instinct, not intellect, that drives the fear of man-made life forms. But it is an instinct shaped by experience. Those such as the Food Standards Agency who exalt science above all things, and who would push it into the vacuum left by religious faith, display a zealot's blindness to the mundane. It was science that told us it was all right to feed animal protein to cattle. Science that killed the rivers and silenced the fields. Science that now accuses GM's opponents of ignorance, prejudice and hysteria. Science that needs to lift its eye from the microscope and take a look outside.

Perhaps I am being too gloomy. Excessive complaint is counter-productive. Through the litany of all that's gone wrong we can blind ourselves to all that's gone right, and make ourselves believe there is nothing left to fight for – or, more insidiously in a world of increasingly centralized decision-making, that there is nothing we can do that will make a difference. We can, and we do. Government is like a diplodocus: you stamp on its tail, and twenty years later the impulse reaches its brain. If you bang on long enough, unless it is George Bush's America, it *will* get the message. At last it sees that there *is* a transport problem; that there is no more room for the waste mountain to grow; that biodiversity is not just the concern of a few farting bean-eaters in love with the nineteenth century; that farm production subsidies in lowland England have been every bit as destructive as logging in Brazil; that global warming will do more than turn Bournemouth into the new St Tropez.

Little and late it may be, but at last we have something that looks like rolling momentum, and all the more reason to keep on shoving. We can hang on to the few good town centres that have not been wrecked – Stamford, Totnes, Bury St Edmunds, Berwick, Beverley, Tewkesbury, Ludlow, Bradford-on-Avon, Saffron Walden, Lewes . . . We can defend the 400,000 miles (650,000km) of hedgerows that still survive. Given an average width of two metres, that grosses up to more than 247,000 acres (100,000 hectares) of land, much of it of incalculable value to insects, animals and birds. By some estimates, hedgerows support 80 per cent of Britain's woodland birds, 50 per cent of its mammals and 30 per cent of its butterflies. To stand before one of Britain's great landscapes on a clear day, even now, is to experience the kind of spiritual uplift that any-where else would require intercession from gods or drugs. Look west from the Malvern Hills; north towards the Cotswolds from the Oxfordshire Ridgeway; north-east from Mam Tor in the Derbyshire Peaks; north-west across Derwent Water towards Bassenthwaite in the Lakes; north-west from Elgol to the Black Cuillins on Skye; or in any direction from a Dartmoor tor. Nowhere on earth does the conjunction of geological upheaval and human endeavour produce a happier chapter of accidents; nothing begs more insistently to be caught on film or canvas. Like eagles and ospreys, however, they must not blind us to the value of lesser things. A scrap of headland in a field of plough is not waste. Nor is any forgotten scrap of wetland, meadow or wood on the urban fringe; nor any scrap of brown land in a city. In common with every other fragment of material that has a value, we discard such things at our peril. If the notion of 'sustainability' is to be more than a parroted mantra, we have to recognize that avoiding waste – of land, buildings and life as well as of aluminium ring-pulls – is the best investment we can make.

CHAPTER FOUR

Down the Aisle

I N THE SPRING OF 2001 THE LAST OF THE GREAT FOOT-AND-mouth pyres was being built in Devon, the stiffened corpses of slaughtered cattle slotted into place like a huge, obscene parody of a drystone wall. In the same beleaguered county, four farmers' wives had rung the police to beg them to remove their husbands' shotguns.

It was an extreme symptom of an industry already traumatized beyond the reach of pain. Nationwide, 29.8m sheep and 576,000 cattle had been culled. Pig farmers were dropping £7 on every animal they sold. Milk that cost 22p a litre to produce was selling at the farm gate for 19p. Farmers' profits over the previous year had dropped by 28 per cent, and the net income from an average family farm had fallen in five years from a comfortable £80,000 to an unsurvivable £8,000. The farmers' grief, however, was a private affair. The nation, as it so often does, let its eye rest on a tragic headline or two and then went back to writing its shopping list. Life moved on. There was meat on the table, as much of it as anyone could eat. Chicken tikka masala in the chill cabinet, mangetout in the fridge. Beyond the usual round of sleight-of-till, cut-and-raise jugglings by the supermarkets, there were no price hikes. No one went without so much as a mid-morning snack and the checkouts never missed a beat. Farmers? Who needed them? Vegetables winged in from southern Europe and north Africa, chicken from Brazil, pork from Poland, apples from Australia, and all of it cheaper than we could produce at home. The Royal Agricultural Society of England spoke of 'cataclysmic change'.

Heading for the midden was the entire British landscape and the industry that had created it. Coincidentally in the line of fire, and facing severe collateral damage, were the food processing industry, village post offices, rural businesses of every kind (pubs, garages, hotels, you name it), and Napoleon's famously misunderstood 'nation of shopkeepers'. Four malign forces had conspired to put the boot in – Victorian imperialism, Adolf Hitler, the European Union and the retail trade. Government policy, like the final injunction to a broken army, was *sauve qui peut*. Farmers were urged to diversify and learn new skills – running bed and breakfasts, teashops, riding schools, IT consultancies – anything but milk more cows. There was a lot of talk about the need to 'respond to market fluctuations' but not much about 'feeding the nation'. In England alone, farmers and their workers were abandoning the industry at a rate of more than 450 a week, leaving a labour vacuum to be filled by immigrant workers – legal and otherwise – who were kept by gangmasters in the kind of conditions that William Wilberforce thought he had abolished in 1833.

The first myth to be got out of the way is that Britain before the Second World War was a coarsely plentiful sub-paradise entirely self-sufficient in food, though quaint it certainly remained. From the historical record emerges a picture of such smock-and-pitchfork antiquity that it seems closer to Merrie England than to the age of the internet. In 1939 there were still 1.5m working horses on UK farms, and only 6,000 tractors. There were no barley barons, turkey tsars or chicken kings. No notion even that farmers had a responsibility to fill the nation's belly. Sixty per cent of the farmed landscape was still under grass, and most farm incomes depended on growing feed – oats and hay – for the horses. There were very few arable fields but plenty of livestock, and a lot of land was given over to horse production. There was no particular need to grow food because, like the Victorians who had shaped our attitudes, we had been able to import all we needed from the empire.

When war came, Hitler could hardly have caused more trauma to the English landscape if he had blitzed the fields directly. With U-boats stalking the sea, imported food was no longer an option. If we wanted to eat, we would have to grow our own. The Spitfire would take the glory, but in terms of self-preservation the most important weapon was the plough. By converting grazing land to arable, even in 1939 you could increase the food value from 100,000 calories per acre to more than 2m – in other words you could feed twenty times as many people.

To speed the plough, the government wielded the tool that would see Britain not only through the war but right through the rest of the twentieth century – the production subsidy. For every acre of grass a farmer ploughed, the government would pay £2 – approximately double the actual cost. More than that: it would pay him a *guaranteed* price for his crop. If the market price fell below the guarantee, then the government would make up the difference. This was the price of wartime self-sufficiency, and no one has argued that it was not worth paying.

But the subsidies, or 'deficiency payments', did not stop when the war ended. World prices could go up and down as much as they liked, but it was all the same to the farmer: he remained as warmly insulated against economic chill as a lamb in a byre. And so it was that the food family became dysfunctional. Out in the fields, the primary producer was so distant from the consumer, and so sheltered from market forces, that he had no influence on, and little interest in, what was being charged for his produce in the shops.

After 1945 the world was impatient to move on. The new battle was against rural tradition; against the limitations of an ancient technology that had evolved over 10,000 years; against the obstinate slowness of nature itself. What grew when a farmer simply drilled seed into the ground was no longer enough. Where once the fertility of the soil had been preserved by craftsmanship – feeding and cleansing the land by crop rotation and animal fertilizer – now it was preserved by science. Out went the horse. In came mechanization, with each new generation of machines bigger and more powerful than the last. Out went the nursery-book mixed farm with its patchwork of fields and meadows. In came monoculture and boardroom management. Food began as it ended, with packeted convenience – pesticides and growth-promoters at one end; microwave dinners at the other. For a while, farm incomes soared.

And then came Europe. When deficiency payments ended, it was only to be replaced by the self-confounding paperchase of the European Union's Common Agricultural Policy with its pounds-per-tonne production subsidies, intervention boards, butter mountains, wine lakes and – the very last stop on the line to meltdown – set-aside, with farmers being paid to grow weeds. The vicious cycle turned faster and faster. Higher production, lower prices, higher production. The system was no longer feeding the nation; it was feeding itself – feeding *upon* itself. Even the subsidy system was self-cancelling and contradictory. While some farmers were claiming extra payments for removing hedges, ponds and walls, others were being paid to put them back again. While Europe fiddled and Maff muddled, the industry burned.

Globally, the flames were fanned by economic collapses in Russia and Asia that wobbled the world economy. While demand shrank, the production express went on racing towards the buffers. BSE meanwhile had done its bit to make England the last place on earth from which anyone would buy a farm product – a fatal handicap at a time when world prices were in a deep and prolonged slump – and the pretence that rural England was in safe hands could no longer be sustained. The unavoidable question – was the desecration of the countryside a fair price to pay for 'cheap' food? – increasingly invited a negative answer. Swine fever in East Anglia brought yet more horrific images to our screens, until the only truly strong and healthy animal left in the national bestiary was its currency, the indefatigable pound sterling itself.

For farmers, it was like being savaged by their own dog. The tragedy was that their subsidies under the CAP were paid in *euros*. The high pound therefore depressed the value of their subsidies just as certainly as it had depressed their exports and rolled out the red carpet for cheaper imports. The government stepped in with 'agrimonetary' adjustments to help bridge the gap, but there were too many holes in the dyke for one finger to plug. By the spring of 2000, total income from UK farming was a mere 35 per cent of what it had been in the mid 1970s. While British consumers were simultaneously spending £3 billion annually on agricultural support and paying the world's highest food prices, the nation's farmers were unable to earn their keep.

The one thing you could not accuse them of was inefficiency. When it came to wringing the last bushel of grain from a hectare of wheat, or the last drop of milk from a cow, they were – and remain – very, very good at what they do. Their productivity has been going up for decades. In 1945 a hectare of wheat produced 2.6 tonnes of grain. In 2000 it produced 8.01. In 1945 the UK had 2.7m dairy cows. By the turn of the century the number had actually fallen, to 2.1m, yet the overall yield had nearly doubled – from 8.1 billion litres of milk in 1945 to 14 billion in 2000. But here came the catch. As productivity climbed remorselessly towards the ceiling, so the profitability graph fell towards the floor. It was the unyielding cycle of market economics – supply and demand. Greater productivity meant increased supply. Increased supply meant lower prices. Lower prices meant lower incomes. Lower incomes meant that, to make ends meet, farmers had to grow even more. And so the cycle went on. The harder they worked, the worse they suffered. By 2002, agriculture's share of the national economy was down to just 0.9 per cent, and a smaller and smaller proportion of retail food prices was trickling back down the line. In the view of the Curry Commission (see p. 162), this was due only in part to the value added by processing into ready meals. Just as important, it suggested, was 'weak selling and farmers' lack of influence in the market'.

And this explains a lot. You begin to see, for example, why

it was that corners might have been cut. Why space-wasting hedges, ponds and copses were swept away. Why growth-promoting chemicals were, and still are, pumped into everything from chicken to barley. Why lowland England bled half to death on the altar of maximum yield. Even so, the CAP was not the only threat to life as we knew it. Another force had assembled, multiplied and hemmed us in – a force far more single-minded, more ruthless and less given to sentiment than anything even remotely answerable to an electorate. In the event, we were a pushover. Like expert judokas, the super-markets absorbed our energy, rolled with it and turned our own power against us. They saw that we were greedy, lazy and materialistic. We valued convenience above quality, con-sistency above variety, and cheapness above heaven. We might not enjoy all the consequences of the choices we would make, but the supermarkets knew that short-term self-interest would tip the balance. Offered the one-stop shop, we couldn't wait to grab a trolley. One way or another the impacts would be felt not just in the fields, farmhouses and high streets of the United Kingdom but in African villages and fragile communities throughout the developing world. At home, our weakness for a bargain created a monster whose fire would leave no corner of England unscorched.

The key date was 1964, when the government abolished the right of manufacturers to dictate the prices of their products in the shops. The end of Retail Price Maintenance, as it was called, looked like the best news for customers since the end of wartime rationing. No more price fixing! Now there would be real competition; lower prices; better value; a genuine shift of power. There was a shift all right, but that word 'competition' hardly fits the bloodbath that followed any better than it would have suited the battle of the Somme. The new supermarkets opened fire with loss-leaders; then own-brands; then buy-one-get-one-free; then loyalty cards, banking services and an expansion from food into every corner of the household market from greetings cards to televisions, spring bulbs, clothing, alcohol, tobacco, newspapers, videos, petrol – anything they could identify a market for, they sold. And whoever had sold them previously – corner shops, bakers, butchers, greengrocers, post offices, hardware stores, pubs, newsagents, sweet shops, filling stations, florists, off-licences, delis, dairies, pharmacies – were left to play spot-the-customer. It was not just the near neighbours who suffered – in Bedfordshire, a shop ten miles from a new supermarket reported a 75 per cent drop in turnover. Between 1976 and 2001, more than 60,000 small retail businesses were felled by the withering fire of the heavy brigade. But even in a battle-field heaped with corpses, no truce will be called. Manchester School of Management predicts that independent food stores will be entirely extinct by 2050.

The rate of decline is progressive, like a whirlpool sucking us into a vortex. As more and more close, there comes a time when it is no longer possible for even the most loyal customers to buy all they need from local shops. The switch to the super-market then rapidly accelerates and the local economy is plunged into freefall. Supermarkets swell into superstores and bankrupt whole high streets; superstores fatten into hyper-markets and obliterate entire districts. Seventy-eight per cent of English villages have already lost their shops, and rural post offices are as rare as corncrakes. With the shops go the other local businesses – builders, decorators, electricians, window-cleaners, carpenters, accountants, solicitors – that depended on them, as well as the local producers who kept them supplied. Colin Breed MP, in a report for the Liberal Democrats, quoted the example of sandwiches sold by a baker in Aldeburgh, Suffolk. The lettuce came from a neighbouring greengrocer who bought it from a local wholesaler who was supplied by local growers. Other local farmers and wholesalers produced the meat, which the baker bought from three different butchers.

Symbiotic relationships like these are as much social and cultural as they are economic. They are the leaves and stems of organic local communities, linking people together, making a complex and smooth-running whole out of the many disparate parts. It looks chaotic but – if left alone – is inherently, if unfathomably, stable. Knowing your customers is not just good business; it is part of the common currency of inter-dependent lives. All this, too, is lost when the supermarket comes to town. Local economies have been compared to coral reefs – teeming with life but ultimately fragile, and impossible to recreate after they have been destroyed. Farmers and horticulturalists used to supplying local shops cannot scale up to satisfy the demands of huge companies that think and buy globally. Not even local wholesalers can survive. Super-markets are not interested in open markets with competitive bidding from greengrocers or rival multiples. They want tightly controlled supply lines to which access is denied to all but themselves. Wholesalers and local markets give way to packing stations. Symbiosis is blown away by a brutal power that sees everything sold by a competitor as an affront.

In 1973, the public flotation of Sainsbury was the then

biggest ever on the London Stock Exchange – every share could have been sold forty-five times over. In 2002 the American cut-price giant Wal-Mart, owner of Asda, overtook the oil superpower Exxon Mobil to become the largest company in the world. The power of the UK's top five – Tesco, Sainsbury, Morrison, Asda and Somerfield – is awesome, a ground-trembling confrontation of bellicose giants. The fighting will never stop. Market analysts have long predicted that the process of predation – big dogs eat little dogs, then bigger dogs eat each other – will eventually reduce the Big Five to two or three ultimate megadogs. In early 2004 the balance took another seismic shift when, after a lengthy stand-off and siege, Safeway fell to Morrison.

Competition legislation will prevent any of them achieving a national monopoly – legally defined as 25 per cent of the market – but many of them (not just the Big Five) already enjoy area monopolies with local market shares well above 50 per cent. Ninety-five per cent of people in the UK now do their main shopping at supermarkets. From such a position of strength, dealing with the opposition is not so much arm-wrestling as fly-swatting. No corner shop can stand up to a heavily armed multiple able to sell what are known in the trade as KVIs – 'known-value-items' such as bread and milk – at less than cost. If this creates food deserts in which non-car-owners – typically the elderly and the poor – have no access to shops, then this is a small price to pay for the overwhelming advantages of competitive trading . . . Except that the competitiveness is illusory. Loss-leaders camouflage the fact that most other prices in the store are not discounted in parallel, and are shielded from competitive pressure. In many instances, particularly in the fresh produce aisle, the prices will be higher than those in the surviving neighbourhood shops. (Friends of the Earth found, for example, that supermarket prices for Cox's apples were 32 per cent higher than market stalls'.) Supermarkets reap the benefits of what trading law calls a 'complex monopoly'. There is no formal collusion between the companies but a broad uniformity in pricing and little real competition. Monopolistic power also enables the practice of 'price-flexing', whereby supermarkets charge different prices for identical products in different parts of the country. As local competition weakens, so the dominant multiple is able to push up the prices.

There is a similar imbalance in the supermarkets' relationships with their suppliers. Growers, packagers, processors and manufacturers are racked and thumbscrewed by retail superpowers that can make or break them overnight. If Al Capone had been into groceries, the outcome would have been much the same. No uptown protection racket was ever more ruthless. Often, suppliers are forced to pay supermarkets simply for the right to go on trading. They may then find that their prices are cut, retrospectively and without consultation, even after they have signed a contract. Entire crops may be rejected if they don't conform to arbitrary standards of colour, shape and size. Further payments may be demanded for inclusion in special promotions. Suppliers may be asked to meet the cost of dealing with customer complaints, regardless of blame, and may be forced to use transport companies or printers who are themselves paying the supermarkets for the privilege of staying in business. And always there is the unforgiving, year-round pressure on prices. As one vegetable grower put it: 'They'll give you a good price the first year. Then the price goes down each year until you quit. They just go somewhere else.'

Various attempts have been made to curb the supermarkets' power, but nothing so far has worked. In 1993 the Conservative government, surveying the wreckage of what once had been the nation's high streets, repented of its market-driven free-for-all and tightened the restrictions on out-of-town superstores. But it was too late. Too many of them already were on the loose, chewing holes in local economies, pushing up traffic congestion and pollution, making possession of a car more important than ownership of a shopping basket. Townspeople paid for superstore convenience with their livelihoods. In 1998 the British Retail Planning Forum concluded that every major supermarket branch caused a net loss of 276 local jobs. Profit leaked away too. Cash spent in local shops selling local produce is vastly more valuable to the community than the same amount spent in a chainstore. The longer money stays within the neighbourhood economy – the more often it changes hands locally – the greater the community benefit. When a supermarket opens and local businesses close, the cash flow is interrupted. Instead of circulating locally, customers' money is diverted to head office salaries, administration and worldwide distribution costs and shareholders' dividends. In a peculiar reversal of market polarities, in this context smaller is bigger. The New Economics Foundation calculated that £10 spent on an organic box scheme would generate £25-worth of benefit to the local economy – £11 more than the same amount spent in a supermarket.

In October 2000, the Competition Commission published a report on 'the supply of groceries from multiple stores in the United Kingdom'. It was a lengthy document, cloaked in caveats and backward-leaning in its anxiety not to be over-critical, but its conclusions were the same as everyone else's. Loss-leaders were against the public interest. So was price-flexing. So was the coercion of suppliers, which the commission decided was damaging to both quality and choice of food. Loss-leaders and price-flexing it concluded were beyond the scope of its control, but it was determined to do something about the supply chain. What was needed, it said, was an enforceable Code of Practice. The job of negotiating this fell to the Office of Fair Trading, which signed up the then 'Big Four' – Sainsbury, Tesco, Asda and Safeway, each of which controlled more than 8 per cent of the UK market – to a voluntary agreement due to begin on 17 March 2002. It was supposed to throw a firewall around the suppliers and stop the rough handling. Out would go the instruments of torture. Deals would be struck and kept through negotiation, not arm-twisting; fairness, not muscle power, would be the guiding principle. Under the code, supermarkets were to set out their terms of business clearly and in writing. They were not unreasonably to delay payment. There were to be no retrospective price cuts; no passing on of supermarkets' marketing costs; no compensation expected from suppliers if supermarkets failed to meet profit targets; no payment required from suppliers to cover the cost of wastage or consumer complaints that were not the suppliers' fault. Only in clearly defined circumstances could payments be demanded as a condition of trading, or for more prominent shelf display; and there were to be no changes to product specifications without 'reasonable notice'. On the face of it, the only undeserving losers would be makers of sleeping pills. Freed from their nightmares, farmers, growers and processors henceforth could sleep like babies.

If anyone ever actually believed this, it did not take long for their optimism to drain away. Throughout 2003 the OFT reviewed and monitored the code in operation, and in February 2004 it published its conclusions. The supermarkets were bullish. The code had simply proved what they had always known – that the system was a model of fairness and equity from which both sides benefited. So who needed it? In Tesco's view, the code 'had not resulted in any significant change in relationship with suppliers'. On the contrary, it had only 'formalised existing good practice'. Not a single supplier had complained to it about breaches of the code. Its closest competitor, Sainsbury, took the same line. It, too, insisted that the code had made little difference to its relationships with suppliers, and had 'merely codified existing practice'. It, too, had received no complaints. Safeway and Asda echoed the same line. Safeway did admit to one complaint – from Express Dairies, which claimed the supermarket had withdrawn its business without giving due notice – but the complaint pre-dated the code and so did not count as a breach. Asda said it had received two complaints, both of which had been settled without recourse to mediation, and both of which, in the super-market's view, were pursued by the suppliers as a 'negotiating tactic'. From the boardrooms came the soft sound of haloes being buffed.

From the suppliers, however, came howls of pain. The code did not work; there had been no improvement in supermarket behaviour, and the proof of this was perverse: it lay in the very absence of complaint. Seventy-three per cent of its respondents told the OFT that the real reason for the suppliers' silence was not the purring state of contentment the super-markets had suggested. It was fear. Such was their terror of being dropped, or 'blown out', if they made a fuss, they had continued to suffer every injustice the supermarkets imposed on them. And it was not just the minnows that were taking a beating. Fear gripped major food manufacturers just as much as it did the small-scale fruit or vegetable growers. All depended on supermarket patronage; all could be fatally damaged by loss or alteration of their contracts; none dared complain. So deep was their dread that they would not allow their trade associations to raise objections on their behalf, or permit anything to be said in public that might identify them. According to the OFT: 'Four of the trade association respondents and three individual supplier respondents . . . refused to allow us to reveal their names or even the sectors in which they are involved.' In these circumstances, as the OFT itself acknowledged, the code was useless. 'While we recognise the fear among suppliers, there is little that can be done under the current code, or indeed any code of this nature, however rigorously drafted, if suppliers are not prepared to assert their rights under it.'

It was a classic culture clash, of liberal reformers taking on the fundamentalists. The code's worst weakness was its own moderate call to reason. Payments to suppliers, it said, must be made 'within a reasonable time'. Supermarkets could not demand price cuts without 'reasonable notice'. Nor should they 'unreasonably require' a supplier to contribute to marketing costs, or charge lump-sum payments that did not represent a 'reasonable estimate' by a supermarket of the risk it took in displaying a new product. And so on. But who was to decide what was 'reasonable'? The OFT conceded that this vagueness of terminology, in which its adjudications would have to rest on the subjective, and thus unpredictable, interpretation of an indefinable concept, was in itself a deterrent to complaint. Thus its report in part became an elegy for its own good intentions:

> ... the mediation process in the code was meant to be an important part of it. It was anticipated that the consideration of complaints heard by the mediator should, over time, lead to the establishment of precedents on the meaning of reasonableness and that this would, in due course, lead to easier and faster mediations and resolutions of complaints and thence to better contractual relationships between the supermarkets and their suppliers. This process has not even begun to happen given the climate of fear among suppliers when dealing with super-markets. The supermarkets appear to be left to decide, with little interference from suppliers, what is reasonable in a trad-ing relationship.

What the Big Four deemed reasonable was clearly indicated by what the review uncovered. Suppliers from the grocery trade's every distant province – packaged foods, drinks, household goods, fresh produce, meat and eggs – all protested that their torments had continued unabated. The commonest abuses remained late payment; retrospective price cuts; enforced contributions to marketing costs and the costs of consumer complaints; lump sum payments as a condition of supply; and contracts made conditional upon the involve-ment of specific third parties such as haulage companies and printers. The evidence was persuasive but ultimately useless. Because of the suppliers' unwillingness to be named, it was anecdotal and the allegations could not be proved. In the absence of formal complaint, the OFT had nothing it could act upon. So how could it get itself out of the bind? Unable to rely on the alleged victims, it realized it would have to find evidence of its own. Its solution was what it called a 'compliance audit', during which it would examine a random sample of contracts between the Big Four and a cross-section of their suppliers, large and small. (Such is the scale of the business that a supplier stays 'small' until its annual sales to supermarkets exceed £73m.) This was a big step forward but it begged the question, even so, of whether suppliers would co-operate with their would-be saviour. As one unidentified trade association put it: 'The code as it stands is practically useless, particularly for small suppliers who cannot afford to lose the investment they have to make to gain the business in the first place. Hence the extreme reluctance to jeopardise their position by taking the customer [the supermarket] to arbitration.' The OFT report on the compliance audit was promised by the end of 2004, but was delayed until 'early 2005' (see Afterword).

In reality, the code is less a judicial device than a gauge of market pressure. Even if it was honoured in every particular, the market would still apply the same irresistible force – downward on price; upward on profit. Code compliance will not stick the farmed landscape back together again; will not replant the orchards or restore life to silent fields. Top fruit have been classic victims of a regime that demands economies of scale, and that puts convenience and consistency above variety and flavour, never mind the rural economy. The story of England's apples is a tragedy without parallel, the names of traditional varieties now so long forgotten, and so improbable-sounding, that you could mistake them for a joke. How many people have sunk their teeth into a Kitchen Door, or would imagine a Catshead to be other than a feline braincase? Everyone, on the other hand, knows everything that's worth knowing about Granny Smith and Golden Delicious.

Cox's Orange Pippin is, or ought to be, one of the most complex mouthfuls that ever fell from a tree. Attempts to describe it fall headlong into winespeak – honey, nuts, peardrops and a shipload of oriental spices have been adduced among its flavours. But who now can remember what it is supposed to taste like? The supermarket version is like a postcard reproduction of a masterpiece – a reminder of the original, but not a substitute for it. Though its pedigree is wholly and typically English – it was raised in the 1820s by a retired brewer, Richard Cox, in his garden near Slough – much of what we buy now is shipped from New Zealand. The nurseryman-connoisseur Edward Bunyard, in *The Anatomy of Dessert* in 1929, had no more fear of hyperbole than he had of greenfly. For him, a mature Cox was 'the Chateau Yquem of apples'. Being an epicure, he had probably never insulted his palate with anything that could suggest a metaphor for the modern industrial Cox, numb from the cold-store. The Pepsi-Cola of apples? The Lucozade? As bland leaders, modern commercial apples march shoulder to shoulder with Bisto gravy, Mother's Pride, tinned lager and industrial cheddar. 'The housewife', that destructive and sinister succubus invoked to justify every loss of quality or choice, wants 'crunch'. She wants to know what she's getting. She wants regular size and colour – a nice fresh green, perhaps with the faintest blush of red. She doesn't want anything pulpy and, with the perverse exception of Golden Delicious, she doesn't want anything yellow. That's the colour things go when they're past it.

The freshness of a supermarket apple owes more to technology than it does to any recent acquaintance with a tree. The year-round availability of Cox, Gala, Braeburn, Granny Smith, Golden Delicious and the others is due to cold-storage techniques. Controlled atmospheres allow apples to be kept in suspended animation for months beyond their season. For the technique to work, however, the fruit has both to be picked early and sold quickly when it leaves the store. Joan Morgan, in *The Book of Apples*, laconically describes the result. 'Traditionalists,' she says, 'argue that . . . many apples now on sale are simultaneously unripe and past their best.'

Nevertheless, the apples are crisp, of production-line regularity and available in the kind of tonnages that keep unit costs down and accountants dancing on their desks. Slicked-down production, storage and marketing have reduced the range of varieties offered in British supermarkets to a small cast of global heavyweights, about twenty in all. Contrast this genetic impoverishment with the National Apple Collection, run by the Brogdale Horticultural Trust at Faversham in Kent, which grows more than 2,300 different varieties – crabs, eaters, cookers, cider apples and decoratives. Not all of them are English (it is a world collection), and not all are blessings to the palate. The French Pomme d'Enfer, for example, the 'apple from hell', is so sour it will pucker your whole head, never mind your mouth; and the local variety, Faversham Creek, has no obvious virtue save a unique ability to thrive in estuarine mud. But every great cellar contains a few sour bottles. Among Brogdale's enormous gene lake are flavours, textures and colours of such range and subtlety that comparisons with viniculture seem to flatter the grape rather than the apple. Naming a South African Cox as your favourite variety would be like naming Blue Nun as your favourite wine when you haven't tasted Sancerre. Viewed from the air, the Brogdale orchards look like chunky ribbed knitting – pink in spring, green through the summer, carelessly paint-splashed red, russet, yellow and gold in autumn and early winter. On the ground, every label is full of mystery and promise. What do we know of Farmer's Glory, or the Transylvanian Fekete Tanyeralma, or the Czech Misen Jaromerska, or Minnesota Russet, or Peggy's Pride, or Pig's Nose Pippin, or Peasgood's Nonsuch, or Norfolk Beefing, or Scotch Dumpling, or Sweet Caroline?

Sometimes a name offers a clue. Fraise de Buhler has the flavour of strawberries. Pine Apple Russet of Devon, like Ananas Reinette, tastes of pineapple. Fenouillet de Ribours hints at fennel. Golden Melon lives up to its name. So do Gooseberry, Green Balsam, Honey Pippin, Irish Peach, King Coffee, Lawyer Nutmeg, Lemon Pippin, Lowland Raspberry, Martini, No Pip, Twenty Ounce, Washington Strawberry, Violette and Winter Banana. Other names explore the subtleties of shape – Catshead, Sheep's Snout, Brown Snout. Others recall the names of their discoverers, or their birthplace, or their pomological ancestry. Scholars of the apple classify them in nine different shapes, from flat (Devonshire Quarrendon, for example) to conical (Adam's Pearmain). Colours range from palest buttermilk to deep burgundy, with an infinity of variations on the theme of stripes, freckles, blooms and blushes. The skin itself may be gossamer thin or Michelin thick. The texture of the flesh may be as soft and melting as sun-warmed slush, or so hard that every bite rings like a gunshot. Cooked, it may melt to foam or hold its shape like turnip. Trying to describe how each variety tastes is like trying to describe colour, or fragrance, or music. You run out of words long before you run out of flavours. In early season, July and August, you are in the realms of, say, fresh young Muscadet – the flavours are crisp, green, perhaps a bit un-finished and gawky, with hints of angle-iron. By the time Christmas comes round, you've run the gamut of acid drops, fresh fruit cocktails, spices and floral accents, and are thinking of old armchairs, Madeira, marrowbones and tawny port. The early varieties – Beauty of Bath, for example – have no stay-ing power. They practically rot as they drop, and must be eaten straight from the tree. Later varieties like the Essex favourite D'Arcy Spice, can be laid down like fine vintages to mature all the way through until Easter. Even a good Cox needs a couple of weeks to find its form.

How has it happened, then, that in a land of such plenty we have so compliantly settled for dearth? Why, when we might fill our bowls with Egremont Russet, Blenheim Orange, Ribston Pippin or Ashmead's Kernel, do we even *think* of Granny Smith? Why are so many of the world's finest varieties reduced to gene samples, hanging on by their root-tips? How could such waste be contemplated, let alone encouraged or condoned? If the apple in the Garden of Eden prompted the Fall of Man, then man in return has ensured the fall of the apple. Its entire history, from primeval crab to supermarket cold-store, is set out in Joan Morgan's *The Book of Apples*, in which she also catalogues and describes more than 2,000 surviving varieties. Britain's pomological love affair has ancient and mystical origins. Celtic tribesmen invested the native crab with powers of love and fertility, and celebrated it

142

in rituals that would later decline into the folksy customs of wassailing and apple-bobbing. For a civilized interlude, the Romans introduced some sweeter, domesticated varieties, though their orchards did not long survive the collapse of empire. The Normans added a boozy passion for cider, and the medieval English, for whom all the world was a drugstore, recognized in the ferocious sourness of the local fruit all the hallmarks of an efficacious medicine. Its reputation in this context may well have been enhanced by relieved constipation sufferers, though its prescription for gonorrhoea may have led more often to disappointment.

It is thought that Britain's first named variety was the Pearmain, adopted for cider-making in the thirteenth century. It was like the first kiss at an orgy. By 1629 the royal botanist John Parkinson was able to list sixty recommended varieties, including Catshead, Pound Royall, Kentish Codling and half a dozen pippins. Promiscuity ran riot, and by the end of the eighteenth century there was chaos. The number of apparent varieties by now ran into thousands, but many of them were useless, bastardized midgets or deceiving lookalikes. 'As a result,' says Joan Morgan, 'a particular apple was often known by a different name not only in another country or county, but in an adjacent village; old varieties were introduced under a different name and claimed as new improvements; and inferior but similar-looking apples might masquerade as an esteemed variety.' (As recently as 1971, the official list of classified apples contained 40,000 names for only 2,000 distinct varieties. Blenheim Orange alone had 62 different synonyms.)

It fell to the London (later Royal) Horticultural Society to sweep the orchard clean in the nineteenth century – 1,396 varieties were authenticated by 1831. No dinner party in

Victorian England was truly grand without a looming pyramid of spotless, estate-grown fruit, with whiskered gentlemen holding forth over their Ribstons with all the wordy passion of wine-bores. Yet, even as the juice was dribbling through their beards, the English apple was plunging into crisis. It was all very well for the gentry to savour their grand vintages – the commercial market was a different matter altogether. In the fields of Kent, the language of connoisseurship might as well have been Ancient Greek. Faced with a choice between the spotty, crab-like runts of the English commercial orchards and cheap but immaculate imports from France, Canada and America, 'the housewife' was voting with her purse. The fight-back was immediate. 'Connoisseurs, enthusiasts, landowners, head gardeners, farmers, nurserymen and market gardeners,' says Joan Morgan, 'all rallied to mount a veritable "fruit crusade" to modernise old orchards, guide the new fruit growers, beat the "Yankies" and persuade fruiterers and their customers to buy English apples before it was too late.' The campaign did have an effect, though not quite the one the connoisseurs had intended. By the late 1890s, Australia, New Zealand and South Africa had joined the fray. Their strategy, deadly effective, was to focus on a small number of reliable varieties that they could ship in bulk.

Local growers quickly learned from their bloody noses. It didn't matter how flavoursome a variety was. If it was difficult to grow, or was susceptible to pests and diseases, or cropped irregularly or bruised easily, then it simply didn't have what it took to beat the invaders. With supreme irony, however, the English growers almost blew their heads off with their own most potent weapon. Cox's Orange Pippin was so plagued by disease that, by 1909, it seemed likely that it would bring the entire industry to ruin. In the nick of time, with American Jonathans and Canadian McIntoshes massing on the horizon, the cavalry arrived in the shape of lime sulphur sprays and the Cox regained its healthy complexion. Partnered by the world's supreme cooker, the Bramley, it survives to this day as the mainstay of the UK industry.

'The housewife' and the supermarket buyer still like their Cox, but in general they find their preference for 'finish' over flavour more easily satisfied from abroad. Imports account for fully 70 per cent of the apples sold in the UK market, which is becoming ever more competitive – even China now is a force in the fresh produce aisles. As with all commodities, it is the supermarkets that control the market. As with all commodities, they will buy from wherever in the world a reliable supply can be found at the lowest price. If New Zealand beats Kent, then no tears will be shed for the one-time Garden of England. For UK growers the results have been catastrophic. In 1945 Britain still had some 80,000 hectares of orchard and the fruit-picking season in the rural calendar was almost as important as the corn harvest. Now just 20,000 hectares remain, and the rate of loss continues to accelerate. In 1987 there were still 1,500 registered fruit growers in Britain. By 2004 only 500 were still in business, and traditional fruit-growing counties such as Essex and Kent had lost 90 per cent of their trees.

It is not just a problem of competition. The disinclination of supermarkets to accept what nature provides, with all its infinite variations of shape, colour and size, means that fruit has to pass a beauty parade rather than a taste test – and much of it fails. Only twelve out of a hundred growers surveyed by Friends of the Earth in 2002 achieved an acceptance rate of 80 per cent or more. Others had seen half their crop rejected;

some had failed to sell as much as a single apple. Some of the rejects had been sold, at a loss, for processing; some were not even harvested but left on the trees, eventually to rot on the orchard floor, or dumped. Almost all the waste was of good eatable quality – not the kind that would have excited a Victorian connoisseur, but no worse to eat than any of the stuff that did make it on to the shelves.

The most common reason for rejection was harmless minor blemishing, or natural 'russeting', on the skins of perfectly sound fruit. One wonders how well supermarket buyers themselves might stand up to such scrutiny. Would possession, say, of a mole make them taste less good? (The threat of rejection is not as improbable as it sounds – beard-wearers already risk disqualification as supermarket job applicants.) Some apples and pears were rejected for being the 'wrong' size or shape. Some for having too much colour and some for too little. Bramleys were even turned away for showing their character-istic blush of red – 'cookers' in supermarkets are 100 per cent green, never mind what happens on the tree. The madness does not stop with rejecting the natural skin colour of a variety which, in any case, is almost always peeled before use. Achieving cosmetic perfection, often in varieties that are not well suited to local growing conditions, means that growers have to apply more chemical controls, with the risk of leaving more pesticide residues in the fruit. Respondents to the FoE survey also reported the full range of contractual malpractices – supermarkets deciding no longer to stock particular varieties, or making last-minute adjustments to the specification of the fruit, or changing the packaging requirements. Whatever the whim – firmer, softer, bigger, smaller, redder, greener, rounder, flatter – the result is always the same. More fruit goes to waste.

The supermarkets' obsession with uniformity, blandness and ease of handling strikes hard at every kind of regional variety. It is commonplace – almost obligatory in Britain – to denounce the French for their local protectionism. Yet simultaneously, and without apparent irony, we admire, celebrate and participate in their success every time we reach for the cheeseboard. France, like Italy, rejoices in the regional variety of its produce and resists the stupefying homogeneity of standard product lines manufactured to supermarket specifications. The dairy industry in Britain has had an even harder time of it than the fruit growers, though for rather different reasons. Its problems began in the First World War. Many businesses died with their proprietors in the trenches, and others perished in the depression that followed. The imposition of statutory control by the Milk Marketing Board in 1933 applied some much-needed lubrication to the rusted cogs of production and supply, but stability was short-lived. In January of the same year Germany elected a new Chancellor and unknowingly paved the way for a near-fatal attack on English cheese.

When war came in 1939, the Ministry of Food ordered that all milk for cheese should be diverted from the farmhouses and used in factories to make bulk supplies of cheddar and other hard varieties such as Cheshire and Leicester. The policy was well-meaning but devastating. Centralized production cut centuries-old regional supply chains. Instead of easily manageable cheeses leaving the farms on short local journeys, vast quantities of cumbersome raw milk lumbered off into the distance at enormous cost in wear-and-tear, labour and fuel. For every pound of cheese, the ministry had to shift a whole gallon of milk. By the time the war ended, English soft cheese, and farmhouse cheesemaking in general, had suffered a mortal blow. Where once there had been 15,000 cheesemakers there were now just 126, leaving the market nicely softened up for its next three murderous assailants, the supermarkets, the European Union and the Department of Health.

The Milk Marketing Board had little sympathy for small cheesemakers. Monopolistic power enabled it to raise its prices beyond the value of their finished cheese, or bluntly refuse to supply them. For makers of 'territorial' cheeses, which depended for their character on the particular quality of local grazing and small variations in climate, the situation was especially hopeless. The milk supplied by the MMB could – and did – come from absolutely anywhere. A few small makers milking their own herds managed to struggle on. Others bowed to the inevitable and went into bed-and-breakfast. Dazzled by the protectionist pyrotechnics illuminating the night sky over Brussels, the MMB wasn't going to mess about with batty little cheesemakers when its own manufacturing arm, Dairy Crest, could more profitably convert the milk into another few cubic metres of the European butter mountain.

The screw turned again in 1984 when Europe imposed national milk quotas, with each member state being set a production limit based on past yield. Given that much of the shortfall in British dairy production historically had been made good by butter and cheese from Australia, Canada and New Zealand – a supply line now constricted by European diktat – problems worsened and cheesemakers faced yet more difficulty in getting what they needed. Quotas meant that the UK was not self-sufficient in milk production, and that the MMB held a monopoly on the supply and delivery of what there was.

The outcome was dire. 'Farmers making cheese in the last few years,' wrote English cheese's greatest champion, Patrick Rance, in *The Great British Cheese Book* in 1982, 'have had to fight against bureaucratic interference calculated to make them feel like absconding criminals.' While all this was going on, and in the name, as always, of 'the housewife', supermarkets were busy squeezing the last few drops of taste and character out of what they obstinately persisted in calling 'cheddar cheese'. In the brave new world of sell 'em cheap, pile 'em high, there was no room in the economic equation for expensively trained staff dispensing personal attention to individual customers.

'Once they decided that it was cheaper to cut and wrap cheese by machine and leave customers to themselves,' wrote Rance, 'the rot had set in. Cheese had to be tough enough in texture and angular enough in shape to cause no problems requiring human understanding to solve them. They must submit, uncrumbling, to the violation of automatic cutting and pre-packing.' In place of carefully made, unpasteurized, cloth-bound cylindrical cheeses with all their inconveniences of maturation and storage, loomed the anodyne monster of block. Designed to spend its entire life encased in plastic, it was rindless, tasteless and immature, but – *Eureka!* – demanded no more skill from counter staff than a toddler might need to place one building block on top of another. The Milk Marketing Board loved it.

Thus was the plain man's lunch of a cheese sandwich, pint of beer and an apple reduced to its nadir. Everything had gone wrong. The bread was a nappy-liner that tasted worse than the bag it was wrapped in; butter had been usurped by a scraping of vegetable grease that hinted at pasteurized vomit; the cheese was an offence to the name it bore; the beer tasted like household cleaner passed through a cappuccino machine; the apple was like someone's first, failed attempt to invent candy floss.

The only thing organic about block cheese was the language used to describe it. Words grew new meanings like Stilton develops mould. The MMB, for example, decided that the proper technical term for a medium-sized factory making industrial cheddar cheese was – what else? – 'farmhouse', while a large industrial plant was a 'creamery'. The tenacious few still struggling to make proper cheese on proper farms were advised therefore that 'farmhouse' was no longer the appropriate appellation for their products. Instead, they should be described as 'traditional'. It was, and remains, a useful distinction, though the MMB itself tried hard to make it meaningless. Its persistently repeated claim, that there was no difference in flavour between traditional cheese and block, served only to confirm that the British dairy industry had fallen into the hands of people who did not know, or who chose not to remember, what real cheese tasted like.

Fortunately outright falsehoods, even when enshrined in the public policy of a state-run monopoly, have a limited shelf life. The public had already made known its dissatisfaction with horrible bread and horrible beer, and the supermarkets and brewers had been forced to abandon their disinformational bluster and give the customers what they wanted. Now cheese, too, began to come under pressure to make itself worth eating again. The theologies of bulk retailing were non-negotiable – stackability was god – but at least the supermarkets began to understand that a meaningful proportion of their 'customer base' was still influenced by its tastebuds. Perhaps block could, after all, be made to taste like cheese as well as look like it. At the same time, they saw the glee with which foreign producers, notably the French, had moved to fill the vacuum. They knew very well that 'Would you like cheese?' at a dinner party foretold the entry of Brie, Camembert or Roquefort, not cheddar, Lancashire or Wensleydale.

Traditional cheesemakers, too, saw an opportunity to raise the profile of their products and began to talk of 'niche markets'. If they felt any sense of astonishment that their once-humble product had been promoted from mousetrap to gourmet's table, they wasted no time worrying about it. Connoisseurs of cheddar emerged and began to speak of floral bouquets, fruitiness, nuttiness, acidity, length and depth of flavour. In the new lexicon of cheese, it was not good enough to speak of cheddar: you had to specify the *terroir* or *château* from which it came. Names like Montgomery, Quicke and Keen began to be dropped in the same breath as Margaux, Rothschild and Lafitte. The quality backlash had begun, and it was shortly to be handed a landmark victory.

Ironically for an organization that had never retreated from the argument that debasement of standards was what its customers wanted, the Milk Marketing Board's nemesis came in the form of the Ultimate Housewife. Margaret Thatcher may have quit the spotlight by 1994, but her loathing of state monopolies was burned into the Conservative government's psyche. The ludicrous milk-waster Dairy Crest was floated on the stock exchange and freed to develop products that bore comparison with real cheese. The MMB itself was replaced in 1994 by the dairy farmers' co-operative Milk Marque, which controlled some 48 per cent of the UK milk supply, involving 18,000 farmers producing 7 billion litres a year, until 1999/2000, when it fell foul of the Competition Commission and was broken up into three smaller regional co-operatives. Not all the cheesemakers' problems disappeared with their main antagonist, but at least it became easier for them to specify which farms their milk should come from, and to produce cheeses of authentic local character. Encouraged by the ever-widening 'niche', new producers began to spring up, extending the UK's repertoire of textures and flavours to something more closely resembling the French.

For the first time since the Ministry of Food surrendered them to the war effort, *soft* cheeses were being made in significant quantity. Alongside (and often in place of) French classics on smart tables appeared the wondrously creamy Wigmore, made from sheep's milk in Berkshire, and the Camembert-like, Jersey-milk Bonchester from Roxburghshire. Hard cheese lovers got the sublime Lincolnshire Poacher, and macho men got Stinking Bishop, a Gloucestershire variety whose flavour would awaken the tastebuds of the dead. Overall, England now has more than 300 varieties of craftsman-made cheese, many of which stand comparison with the best in the world.

Was this, then, the end of the story, the happy conclusion we had all been waiting for? Well, not quite. The old tensions were still simmering away, and it was by no means certain that the right side would win. Supermarkets had improved their ranges, but the shelves were still dominated by block, and block's more passionate enthusiasts continued to argue that rindless, plastic-wrapped, pasteurized factory cheddar tasted every bit as good as a traditional, cylindrical cloth-bound one made from unpasteurized milk. On both sides of the theological divide, 'pasteurization' raises more hackles than any other word in the lexicon of dairying. In cheeses made from raw milk, which include most of the traditional varieties, curdling is begun by naturally occurring bacteria in the milk. In cheese made from pasteurized milk, which includes all block, the natural bacteria and flavour-friendly enzymes are killed by heat treatment and have to be replaced by laboratory-bred artificial 'starter' cultures.

Traditionalists hate pasteurization because it eliminates the subtle variations in flavour which are the hallmarks of connoisseurship and which give each day's make a different character. To supermarkets, this is heresy. They want safe, consistent products that always taste the same and are not going to throw a tantrum if they don't get expert attention. And the only way to ensure that is by pasteurization. Wherever the cultures clash, there is potential for disaster. To many keen cheese-eaters, a bit of blueing on mature cheddar is added value. To supermarkets it means only mouldy, unsaleable cheese. When traditional makers do supply supermarkets, therefore, they tend to send younger, less mature cheese than they would sell to a specialist cheesemonger. Even so, one maker of good traditional cheddar in Somerset went out of business simply because so much of his craftsman-made cheese was junked by supermarkets to which its very uniqueness was anathema.

Others have been hounded to the brink by bureaucracy. There is a widespread misconception that raw-milk cheeses are second only to plague-rats as vectors for disease. In reality, unpasteurized hard cheeses are no more likely than any others to carry listeria or E. coli. Any micro-organisms surviving from the raw milk are killed off by the acidity of the maturing cheese. Contamination, if it occurs, will be the result of poor handling, not of anything that happens in manufacture. And yet public health authorities react with panic when a food-poisoning case is traced to unpasteurized cheese. Or even, sometimes, when it isn't. Notoriously, in 1995 an attempt by officials of South Lanarkshire Council to have the entire stock of Lanark Blue cheese destroyed was overturned in the courts when the cheesemaker, Humphrey Errington, proved that his pungent, Roquefort-style sheep's-milk cheese was no danger to health. The cost to the taxpayer was half a million pounds. Three years later, in May 1998, when a single case of food poisoning was traced to a sample made by an award-winning cheesemaker in Somerset, the Department of Health banned all sales of the firm's products under emergency legislation designed to arrest the spread of major epidemics. Many in the department seemed unlikely to rest until unpasteurized cheese was banned altogether.

Randolph Hodgson, proprietor of Neal's Yard Dairy and a former chairman of the Specialist Cheese Makers Association, is an uncompromising champion of raw-milk cheeses. When I first spoke to him in 1999 he was pessimistic about their future:

> The best cheeses in the world are all unpasteurized, but I see a threat to their continued production. A consensus of food scientists and technologists perceive a threat to public health, and they think it shouldn't be allowed to continue. Food scientists distrust it because it's outside their remit – a product that's made on the farm and not in a factory. They hate the lack of big-company control.

The stalking horse for unpasteurized cheese has been raw (green-top) bottled milk, which the Labour government initially promised to ban. 'There is no reason why consumers should be exposed to this risk,' said the food safety minister, Jeff Rooker, in November 1997. He was confounded in January 1999 when the agriculture minister, Nick Brown, announced not the expected ban but tighter health checks. It was a welcome reprieve, but how long could the line hold? The sale of raw milk is already banned in Scotland, and nothing is more certain than that it will come under renewed attack in England and Wales, where it is restricted to licensed outlets such as farms, catering businesses and milk roundsmen (but not supermarkets). The Food Standards Agency's advisory committee for Wales has recommended a ban, and the FSA is holding a public consultation before making a formal recommendation to the Welsh Assembly. If green-top milk does go, then the next target is bound to be unpasteurized cheese.

It is not just in the UK that pressure has been applied. The US Department of Agriculture (USDA) tried to persuade the Codex Alimentarius Commission, the offshoot of GATT that lays down international food standards, to ban trade in un-pasteurized cheese. That the Americans failed was due largely to the opposition of the French, but even Gallic solidarity cannot be guaranteed for ever. It remains true that 72 per cent of all French cheese producers, and 49 per cent of all varieties whose names are protected by *appellation contrôlée*, use unpasteurized milk. Analysis of total output, however, tells a different story. Bulk producers command such an enormous slice of the market that fully 82 per cent of the cheese made in France is pasteurized, a figure that can only rise. Raw-milk producers can be relied upon to defend their corner for all they are worth, but the worrying fact is that their corner is worth only 18 per cent of the market. France may still be in touch with tradition, but its power base is shrinking.

In the UK, too, none of the commercial big hitters uses raw milk. Much traditional cheddar is now made with pasteurized, as is Stilton. Given that country's addiction to litigation, no one in the US will want to risk making anyone else take an unscheduled visit to the bathroom – thus it is impossible to believe that the USDA will abandon its antagonism to un-pasteurized cheese, though scientific opinion in America is now tending towards a more balanced view.

In the light of all this – the nervy, trigger-happy attitude of the UK health establishment, the hostility of the Americans, the fatalistic surrender to pasteurization of French and British cheesemakers – you might believe that unpasteurized cheese was some kind of mass killer to rival tobacco or the motor car; that each mouthful of Wigmore, Golden Cross or Lincolnshire Poacher should be preceded by a drum roll. But the truth is rather different. Data collected by the Communicable Disease Surveillance Centre showed the exact breakdown, food by food, of all reported cases of food-borne illness between 1992 and 1996. Top of the table, responsible for 33.6 per cent of the total number of cases, were poultry and eggs, followed by red meat and meat products (18.3 per cent) and fish and shellfish (11.3 per cent). Ultra-healthy salads, vegetables and fruit scored 11.2 per cent, and drinking water, as one might expect, only 2.7 per cent. And cheese?

Just 0.1 per cent.

And yet skilled producers of hygienic, well-made cheeses were being treated like pornographers, treading a legal tightrope, their very smallness an offence to food scientists reared on mass production. Like huge factories whose managers needed to reassure themselves that health and safety procedures were being followed by their workers, one-man cheesemaking operations were forced to install Hazard Analysis Critical Control Point (HACCP) systems, involving checklists, tickboxes and timecharts.

Was the milk delivery at the correct temperature? Tick box. Was the hose connected properly? Tick box.

Was the acidity correctly adjusted? Is the salt from an appropriate source? Has the lavatory been cleaned? Did you wash your hands? And on, and on, and on. Tick, tick, tick, tick. While systems like these were essential in large companies where workers needed to be monitored, they invited sledgehammer/nut analogies when imposed on a farmhouse where the only performance the proprietor had to monitor was his own.

'The sheer volume of paperwork is turning them into pen-pushers,' said Randolph Hodgson in 1999. 'Instead of looking after his goats, every ten minutes the farmer has to go and tick a box to prove to himself that he is still doing something he has done all his life. He is being ground down by the hassle.' In the official hierarchy of expertise, he said, 'a food scientist just emerged from university ranks higher than a farmhouse cheesemaker who has been making cheese since 1949.' Is it any wonder that so many of them were throwing in the towel?

But here at least there is some better news. The UK Food Standards Agency, established in April 2000, has significantly softened the official line. It is educating environmental health officers towards a better understanding of unpasteurized cheese, and has introduced a 'workbook' system, much simpler than HACCPs, that encourages producers and environmental health officers to sit down together and

157

rationally analyse their systems. At the same time, the EC has ordered a Europe-wide survey of raw-milk cheese which Randolph Hodgson is confident will deliver the clean bill of health it deserves.

W hy expend so many words on apples and cheese? Because of what they stand for. Because they are the magnified image of all that has gone awry in the English countryside – the war on diversity; the contempt for tradition; the confusion of 'market forces' with every monopoly's enemy, 'choice'; the dead hand of the legislator; the crushing of local economies; the supplanting of excellence by mediocrity; the trashing of good people and the literal rubbishing of their food; the deadening waste and stupidity of it all. And yet . . .

In their commitment to a cause, their tenacious attachment to variety and quality, may be seen perhaps the start of the reverse pendulum swing that everyone, from Defra to the Soil Association, has been looking for. It is the point at which imagination and pragmatism meet. British agriculture for years has looked like a patient on life-support, with no hope of surviving unaided. Why not just pull the plug? Let it go, like we let go coal and steel before it. Let the supermarkets fly in strawberries from California, if that's what they want. Let's think globally, not parochially. Enjoy the best of what the world can offer. Why not let the foreigners, with their cheap labour, get on with the dirty business of actually growing and breeding and shovelling the muck while we – applying the wisdom of emperors – import cheap raw materials and convert them into expensive manufactured foods with huge profit margins? It is an argument with obvious appeal to free-market ideologues, and, despite all they have suffered, farmers can hardly rely on public sympathy. People have not forgotten the arrogance with which they claimed dominion over a countryside they systematically abused, polluted and disfigured. It has been difficult to admire their resistance to public access; their antipathy to wildlife; the cruel inhumanities to chickens, turkeys, calves and breeding sows; their lack of gratitude for decades of public subsidy; the graceless way they take our money and then slam the gate in our face. Bankruptcy? It's too good for them. Farmers of all people must realize they can only reap what they sow.

Even on pragmatic grounds, such arguments are unsustainable. Like it or not, we *need* farmers, and we'll go on needing them for as long as our imaginations can stretch. 'Abroad' is not as stupid as we think. It knows it makes no sense to export bulky raw materials for British processors to profit from. Raw materials are much costlier to ship than packaged foods, and packaged foods earn much more profit. You don't need a first in economics to foresee the likely outcome of a collapse in UK farming. From eastern Europe would come not raw milk or milk powder, but raspberry mousse and Choc-o-Milk. From Brazil and Thailand would come not container-loads of raw chicken but cartons of frozen nuggets. The hard reality is this: if the British farmer goes down the tube, then he'll take the UK food-processing industry with him. And how would we keep our food safe then? It is hard enough to keep track of supplies within the UK. Unforgettably in 2000 a number of processors and supermarkets, including Sainsbury's, Budgens and Shippam's, had to recall products made from condemned chicken diverted from a Derbyshire pet-food plant. With supplies from abroad it is much more difficult. 'Food is usually monitored on exit from the departing country,' says one well-placed market analyst. 'But if you're a third world country and there is foreign currency at stake, are you really going to say this is not fit for export? The UK consumer would have no idea what he was eating.' Indeed, one of the most important factors that make foreign – and particularly non-EU – produce cheaper than our own is that it does not have to meet the same health and welfare standards. Paperwork is routinely inspected at entry ports. But paperwork won't confess to animal welfare abuses, or to drug or pesticide residues (even if it did, the only result would be a polite letter to the embassy or chief veterinary officer of the offending country).

Even if the safety of food imports could be guaranteed, the case for domestic agriculture would be overwhelming. Who else but farmers would tend the landscape? What other kind of land manager is waiting in large enough numbers to take over? Pies come no higher in the sky than the idea that the landscape could be conserved in perpetuity by anything other than a profitable, confident and forward-looking farming industry. But strong passions are engaged, and there is no formed consensus on what 'forward-looking' might actually mean. For some, including food scientists and voices within the UK government, it means more of the same – more 'conventional' high-yield crops and the introduction of GM. Any contrary view is likely to meet accusations of anti-scientific bias, naivety and backwardness. When an article of mine in the *Sunday Times Magazine* questioned the objectivity of the Food Standards Agency's policy on organics, a professor of toxicology at a leading British university responded with a furious letter:

On the one side you have a collection of environmental, organic and consumer associations who for very different reasons want to turn the clock backwards on agriculture and food and who use slogans about agriculture and assertions about quality which on detailed examination have no scientific basis ... On the other side you have the Food Standards Agency that stands for scientific knowledge as the basis of food policy. If we don't use the best available knowledge as a basis for policy then frankly we might as well go back to reading entrails with all the hazards that would involve. The FSA was not set up to be catspaw for special interest groups and articles such as those by Girling either are written in ignorance of the realities of farming and food or because of ideological commitment by the writer.

All the ingredients are there – the implication that support for organics comes only from crackpots who 'want to turn the clock backwards'; the implication that 'scientific knowledge' and 'best available knowledge' are the same thing; that advocates of change must belong to 'special interest groups' with a commitment to 'ideology'. Such are the black arts of propaganda. The reality – the actual, modern, non-backward-looking reality – is that farming can no longer be seen purely in terms of food production. The buzzword, horrid but unavoidable, is 'multifunctionality'. In plain English, this means that such things as the conservation of landscape and wildlife, the development of alternative energy sources and maintaining a clean water supply should count on the credit side of farming's balance sheet. There has to be more to agriculture than growing commodities at lowest cost for global markets, or farming the subsidies. No one, not even the most purblind of village idiots, now believes in a system that pays farmers to grow food that no one wants.

In January 2002 the Policy Commission on the Future of Farming and Food, known as the Curry Commission after its chairman, Sir Donald, presented its report to the prime minister and the secretary of state for environment, food and rural affairs. With supreme optimism it looked forward to a future in which the English countryside once again is varied and attractive, and has regained its diversity and regional character. Everyone in this sunlit upland of the future is basking in the reflected glow of each other's good health; everything bursts with freshness and good intentions. 'In our vision of the future, farmers continue to receive payment from the public purse, but only for public benefits that the public wants and needs ... [They] are rewarded for looking after their land and for providing an attractive countryside ... Rather than being something farmers once did when they could afford to, good land management is now core business.' Customers, too, have passed into a new age of enlightenment. 'Consumers are health-conscious and take a keen interest in what they eat. They know where it has come from. They know how it was produced. Through their purchasing decisions they reflect their concerns and aspirations for the world we live in.'

The final withdrawal of CAP production subsidies will have removed the temptation to over-produce, and – thanks, perhaps, to the new, more discriminating breed of customer – no dark shadow is cast by the retail trade. 'The key objective of public policy should be to reconnect our food and farming industry: to reconnect farming with its market and the rest of the food chain; to reconnect the food chain and the countryside; and to reconnect consumers with what they eat and how it is produced.' We won't have to hang reproductions of Ford Madox Brown's *The Pretty Baa Lambs* on our walls: we'll be living in it. Having dreamed its dream, however, the commission accepted that the road to paradise would be paved with torments. There was no sign that supermarkets would ease their grip. Cheaper imports would continue to flood in, pressing down on prices and cutting deeper into UK farmers' market share. Home cooking more than ever would mean reheating ready meals bought in supermarkets, and fewer and fewer people would cook like their grandmothers from raw ingredients. 'Already,' it said, 'few people in England now have a direct link with the way that their food is produced, and a knowledge gap is growing.'

How, then, were these people going to inhabit the commission's vision? How would they reconnect to the countryside? How was the bloom to be restored to their cheeks? First off, and most urgently, we need to see the end of production subsidies. The long-awaited change to flat-rate area payments, which 'decouple' grant payments from farm yields, began in 2005 and will take eight years to complete. Production subsidies, said Curry, 'divide producers from their market, distort price signals and mask inefficiency'. Abolition will put new sinew into the industry, encourage it to work on its fitness and rekindle its imagination. 'Some farmers,' Curry went on, 'will need to reinvent their businesses in order to become profitable.' This will involve choosing one or more of three basic options – making themselves leaner and more efficient; clawing profit back down the supply chain by adding value to their own products; or diversifying into new markets. Improving efficiency does not just mean becoming better farmers and adopting 'best practice': it means small farmers circling the wagons, pooling their bargaining power – all for one, and one for all – to bulk up their muscle and wring better deals from retailers. The battle they face, however, is not against just market forces. They are stalked by the Office of Fair Trading and its sniffer-dog the Competition Commission, who are likely to rule, as they did in the case of Milk Marque, that large co-operatives are an abuse of market power. As the principal competitors for the supermarkets' business are even larger co-operatives from abroad, this raises yet more questions about the location of goalposts and the contours of the playing field. Curry wanted the competition authorities to soften their line and take account of the wider context, but the issue has yet to be tested and many farmers have been afraid to take Curry's advice. In July 2004, the OFT tried to clarify its position by publishing answers to a set of 'frequently asked questions'. It certainly illuminated the problems. Farmers' co-operatives were perfectly OK, and unlikely to raise 'competition concerns', just so long as they had no significant impact on the market. They were welcome to share machinery and services (accountancy, websites, etc.). Such co-operatives, the OFT said, 'can be good for efficiency and are very unlikely to harm competition, *especially if their scale is not large* [my italics]'. Size matters, too, when farmers get together to negotiate better prices by buying goods in bulk – '. . . competition issues might arise . . . if the group had a very strong position in the buying market'. It matters especially when they form a marketing or sales group. They can sell through a co-operative but, if they achieve a dominant market position, they must not exploit it by enforcing excessively high prices or by trying to force rivals out of business (if books had soundtracks, we would now cue sepulchral laughter from supermarket victims). Price fixing is illegal regardless of size, as is 'market sharing' – i.e., carving up the country and agreeing to keep out of each other's territories (cue more laughter). The real frustration for small farmers is their vulnerability to bullying by customers who have the power to put them out of business, and against whom the law allows them no commensurate power of redress. This is number 14 in the OFT's list of FAQs:

'If employees can collaborate legally to withhold their labour in order to obtain a fairer wage, why can't self-employed farmers and growers collaborate to withhold their produce in order to obtain a fairer price?'

And this is the answer:

'Collusion among undertakings to withhold produce to get a "fairer" price could mean higher prices and potential shortages to the detriment of consumers . . .'

Level playing field? Judge for yourself. The one clear message is that in its protection of 'competition' as the sole guarantor of fairness and value, the OFT ensures the ultimate victory of the strong over the weak – a moral inversion that places hell in the fields, purgatory in the high street and Tesco in the kingdom of heaven.

The other escape route signposted by Curry is 'added value'. This might mean little more than improving the brand image of UK-produced farm food; or it might mean farmers edging their way upward through the supply chain to adopt the roles, and pocket the profits, of food processors and packagers (another incentive for small producers to band together). The NFU's 'Red Tractor' scheme, launched with government backing in 2002, aims to set 'assured' standards for a range of farmed commodities – cereals, fruit and vegetables, meat and dairy products – but is not well understood by the public, does not satisfy welfare campaigners that animals are any better off, and has yet to develop for conventional produce the kind of strong brand image achieved by the Soil Association for organics. Until it does, the logo is unlikely to confer very much in the way of marketable added value. Price-obsessed supermarkets and their customers in any case are unlikely to pay much of a premium for Britishness alone. For that to happen, they need to see unique selling points – rare breeds, traditional varieties, local specialities. As long ago as 1998, the speciality food sector contributed 5 per cent of the total turnover of the UK food and drink industry, a far from negligible £3.6m. It provided 52,000 jobs – 10 per cent of the food industry total – and had an overwhelmingly local focus, with 64 per cent of producers sourcing more than half their ingredients locally and 45 per cent doing more than half their business through local markets. It is conventional wisdom, endorsed by Curry, that building this market offers farmers one of the best opportunities both to add value and to claim a bigger percentage of the retail price.

Opportunities for protecting local products, and for emphasizing their unique character and value, are provided by the European 'Protected Food Names' scheme, whose purpose is very similar to that of the French *appellation contrôlée* for wines and cheeses. They might not in every case be better than their rivals, but the suggestion that they have special value – a value sufficient to justify an official endorsement of authenticity – is a powerful image-builder. Despite the example of the French, and the fact that consumers in Europe have shown themselves willing to pay on average 18 per cent more for a 'protected name' than for ordinary products, UK producers have hardly knocked each other down in the rush to register. Cheeses not unexpectedly are the largest group, with thirteen varieties listed (Beacon Fell traditional Lancashire, Bonchester, Buxton Blue, Dorset Blue, Dovedale, Exmoor Blue, Single Gloucester, Swaledale, Swaledale Ewes', Teviotdale, white and blue Stilton, and West Country farmhouse Cheddar). Next comes fresh meat (Orkney beef and lamb, Scotch beef and lamb, Shetland lamb, Welsh beef and lamb); then ciders and perries (Gloucestershire, Herefordshire and Worcestershire), beers (Newcastle Brown, Kentish ale, Kentish strong ale and Rutland bitter) and fish (Whitstable oysters, Scottish Farmed Salmon and Arbroath Smokies). Add Cornish clotted cream, Jersey Royal potatoes and Traditional Farmfresh Turkey, and that's the lot – a UK contribution of just 36 out of a European total of nearly 700. France by comparison scores 138. Even Greece manages 83.

To Curry, the message came loud and clear. 'We think that the time has come,' the report said, 'for locality food marketing to become mainstream in Britain as it already has in France and elsewhere.' It showed, too, a sophisticated understanding of brand image. 'Modern marketing, with its focus on values and feelings rather than comparisons or statistics, would seem perfectly suited to marketing food produced in some of England's most beautiful countryside.' The potential, for regional foods as well as organics, is to lift them out of the cosy but restrictive comfort zone of 'niche markets' so that they can compete more aggressively for national market share among consumers who are reckoned to be increasingly worried about the health, quality and provenance of what they eat. Local farmers' markets have been a great popular success, but they remain an occasional novelty rather than the reliable, everyday outlet needed to propel local produce into the commercial mainstream.

To this end, Curry proposed that the government-sponsored marketing organization, Food from Britain, which promotes British food abroad, should direct more of its attention to marketing regional foods within the UK itself. This it has done. Its website, listing more than 3,000 local producers of everything from organic beef to alcoholic ice cream, will be an offence only to the kind of sanctimonious bean-eater who identifies the pleasures of the table with moral decay. In a kind of glorious electronic pageant, it gives new substance to the old, rosy-cheeked myth of a land of coarse plenty. As if to confound the sceptics who argue that this kind of thing is a nostalgia-driven, backward step into the age of the farm cart, much of this new abundance can be bought online. The turning-the-clock-back jibe is a staple of the pro-GM, anti-organic lobby. People who have made expensive mistakes, or who have endured catastrophes in their lives, always wish they could transport a bit of hindsight with them back into the past. 'If only I'd known then what I know now . . .' is the hand-wringing plaint of everyone who has ever suffered at the hands of experience. Well, British agriculture does know now. Having scraped itself up from one catastrophe after another, it has no choice but to let hindsight colour its view of the future. It does have options, but they do not include a failure to adapt.

Inevitably it is the small and medium-sized farmers, as numerically strong as they have been commercially weak, who have most to lose or gain. According to Curry, 80 per cent of UK food production now comes from just 25 per cent of the farms, with the top 10 per cent already providing more than half the total. Some 50 per cent of farms are part-time businesses owned by people who have other sources of income, and whose combined contribution to output is less than 3 per cent. And yet they cannot be discounted. In their impact on the landscape, the minnows are every bit as important as the barley barons. It is this very majority of smaller farmers who represent, as Curry puts it, 'the ideological and political basis for agricultural support', and whose love affair with the land brings them closest to the kind of sustainable, environmentally benign and visually attractive land use that most recreational users of the countryside want to see. Some of these will be able to hang on through efficiency savings, or by joining co-operatives, or by serving specialist markets. Others may have to diversify into non-food crops – oil seeds, biofuels, pharmaceuticals – or other kinds of business altogether. For many, in the future as in the past, the lifeline will be the farmers' traditional standby in times of need, bed and breakfast. Increasingly the frying pan will rival the seed drill as the principal engine of profit.

This is not a trivial point. The beauty of the countryside is its most bankable asset. When Gallup in 1996 polled British people to find out what they liked best about living in the UK, the runner-up to the inevitable chart-topper, 'freedom of speech', was 'the British countryside'. In 1998, an estimated 1.25 billion day trips were made into rural England alone. In 2000 the number of overnight stays topped 33m, generating income of more than £4 billion and helping to sustain 380,000 jobs in rural tourism. Real pragmatists do not reject love of landscape as pointless sentimentality. They invest in it. The gripe for farmers has been that their most enduring creation, the landscape itself, has made huge market opportunities for other, peripheral industries while the core business of food production has been salted with strong men's tears. Tourism depends for its very existence on a managed countryside. All the great set-pieces of English landscape – the lakes, the fells, the moors – owe their character to grazing animals. The cost of losing this is evident in the catastrophic impact of foot-and-mouth disease – an estimated loss to the tourist industry of £5.2bn. In a rational world, it cannot be imagined that farmers will go on indefinitely providing services, in food production and landscape management, from which most of the profit goes to others.

If large tracts are not to degenerate into weedy wastelands, as unattractive to tourists as they are barren of animals and crops, then it is essential that CAP reform proceeds quickly and smoothly. As we have seen, production subsidies, the economic bulldozers that have uprooted the hedgerows, filled in the ponds, fouled the rivers and scoured the uplands, are on their way out, to be replaced by direct payments based on the size of the farm rather than the size of the crop, thus relieving the pressure to over-produce. The idea of 'cross compliance', favoured by Curry, Defra and pretty much anyone else who takes a long view, is that subsidies in future should not only be 'decoupled' from production targets, but should be given only in return for what the jargon calls 'environmental goods' – restoring, improving and maintaining the quality of the landscape. 'Resources currently channelled through the CAP,' says Curry, 'have to be refocused on public goods. Instead of paying farmers in a way that encourages them to hit the land hard, we should pay them to look after it.' It is not easy to construct an argument to gainsay this, though there are some within the Treasury and the National Farmers Union who will do their best.

It is important that they fail. Important, too, that the real mullahs of the food trade, the supermarkets, become publicly accountable for the full range of their social, economic and environmental impacts. If a voluntary code doesn't control their behaviour, then – as the Competition Commission first suggested – it must be given the force of law, with meaningful penalties for offenders. Neither should fair dealing be expected only of the superheavyweights in the Big Four, or however many of them are left standing when the smoke clears. What sense does it make for other huge businesses to be excused compliance only because they fail, possibly by the slimmest of margins, to hold an arbitrary percentage of the national market? What is right for Tesco, Sainsbury, Safeway and Asda surely is no less right for Somerfield, Waitrose and Budgens.

A way must be found to hold them to their oft-repeated commitment to local food – 'the next major development in food retailing', as it was described to Curry. Shrinking the distance between field and plate is important not just to revive local economies. The spread of foot-and-mouth disease, for example, was accelerated by the multiple movements of sheep between farmers, dealers and markets. According to the government's chief veterinary officer, animals could make up to eight journeys by road before they reached the farm at which they would be 'finished' before their last journey to the slaughterhouse – itself an average of a hundred miles. If you were a virus wanting to design a system purely for your own benefit, then you would be hard-pressed to come up with a better one than this. The development of global markets, and the routing of supplies through a small number of central distribution hubs, means that the food chain now girdles the earth. The pressure group Corporate Watch calculated that an English 'Sunday lunch' of roast chicken (from Thailand), runner beans (Zambia), carrots (Spain), mangetout (Zimbabwe), potatoes (Italy) and sprouts (UK) would clock up a combined journey of over 26,000 miles. Five veg might be pushing a little hard at the boundaries of credibility (does anyone take government dietary advice quite that seriously?) but you get the point. Next time you're on a motorway, notice the number of refrigerated juggernauts crossing and recrossing each other's paths. Around the M25, when not trapped in the tailbacks they help to create, they circle London like supplycraft around a barren planet. The distances, too, are cosmic. In twenty years the British freight market has grown by 65 per cent, with food and agriculture now accounting for between 30 per cent and 40 per cent of the total, and it is the UK's fastest-growing source of greenhouse gases. Global market: global warming.

Bulk handling is supposed to promote efficiency and, through efficiency, economy. This may be true in terms of cheapness at the checkout. But your grocery bill is only the final magician's flourish at the climax of a long and complex illusion. The over-consumption of non-renewable fossil fuels, the long-term effects of pollution and costs of road congestion, the loss of local character and variety, the shrinking of local economies and death of competition, the waste of cosmetically imperfect fruit and vegetables, the enslavement of suppliers and cost-cutting pressure on the farmed landscape – all of these drag the bottom line much further down the page. And this is before we even begin to think about the volume of garbage such a system produces.

CHAPTER FIVE

Rubbish

MARCH 1944 IS NOT A MONTH MUCH REMARKED UPON BY historians. The Normandy landings had been post-poned so that more landing craft could be built. In the Far East, Field Marshal Slim was making ready to advance into Burma. At home, as it faced another year of war, Britain was digging for victory, eking out its rations and wasting nothing from which a few drops of 'goodness' might yet be squeezed. The word 'pack' still denoted something carried by a soldier on his back and wouldn't be seen dead without its letter 'c'.

For packaging enthusiasts, however (and, yes, such people do exist), it was a Moment in History. On the 27th of that month, a Swedish designer, Ruben Rausing, applied to patent a new kind of milk container made of plastic-coated paper-board that could be filled and sealed at dairies as an unbreakable alternative to glass. The original was triangular in shape, but don't be fooled. The eventual issue, in May 1947, of patent number GB588343 was one of the wellsprings of the modern world. Ruben Rausing's son Hans, enjoying a personal fortune of £4.8bn, for years took it in turns with the Duke of Westminster to be listed as the richest man in the UK, and rare is the household unvisited by any of GB588343's liquid-filled progeny. The much-maligned but ubiquitous paperboard drinks carton – notorious shirt-soaker, milk-squirter and nail-breaker though it may be – made the Rausings' company Tetra Pak (now part of the Swiss-based multinational Tetra Laval) into the packaging world's equivalent of Ford or Boeing. For fruit juice and milk, it's the only way to travel. In 2003, Tetra Pak worldwide produced 105 billion packages – by its own calculation, that is fifteen packs for every man, woman and child on the planet. Stacked one on top of the other they would make a column 11,394 million kilometres high, a distance equivalent to sixteen round trips to the moon. Inside them were not just drinks but ice cream, dry foods, fruit and vegetables, cheese and pet food.

The advantages are considerable. The sealed cartons are hygienic and impervious to spoiling by light or air. Being sterile, they can be kept for months without refrigeration, thus greatly extending their shelf-life and reducing the wastage of what they contain. They are light and efficient (a ratio of 3 per cent packaging to 97 per cent content). They are easy to handle, transport and store. They are mostly made from renewable materials, and are very, very cheap. The company still repeats, as a moral absolute, the guiding principle of its founder – that a good package should save more than it costs.

What even the best package cannot avoid, however, is its appointment with destiny. A hundred and five billion used Tetra Paks is a hundred and five billion items of rubbish. Even crushed flat, they'll be good for a moon trip or two. The statistics of domestic garbage are like that – on a scale that challenges human imagination, but must not defeat it. 'Must not defeat' because, if disaster is to be avoided, the ingenuity of the producer has to be matched by the ultimate disposer. It is a well-travelled statistical groove, but let us run through it one more time. Each person in the UK on average throws away 4.5 times their bodyweight in garbage every year – well over a tonne per family. But that is only the beginning. Another 5.5 times bodyweight is thrown away by those who make or deliver what each of us consumes. If you go deeper into the chain and look at energy consumption and built infrastructure in the UK, the annual total rises to 100 times bodyweight. If you go deeper still and include foreign economies from which goods and services are imported, it doubles again. The true figure, therefore, is nearer 200 times bodyweight, or 20 tonnes each. And it goes on growing. By the government's own estimate, the volume of municipal waste rises by an average of 3.2 per cent a year – faster than GDP, and enough to double the flow in less than twenty-five years. Between 1996/7 and 2001/2, it rose by 17 per cent. The most famous statistic, repeated ad nauseam, is that the UK *every hour* discards enough waste to fill the Albert Hall – a literally unimaginable 430m tonnes a year. Of this, municipal waste – the stuff collected by local authorities from homes, shops, schools and small businesses – accounts for just 7 per cent. The other 93 per cent is from quarrying (27 per cent), agriculture (20 per cent), demolition and construction (19 per cent), industry (19 per cent) and dredging (8 per cent). All of this has to find either a last resting place or a new use. In the statistical broth, with every kind of figure stirred into the same pot, there is persistent cross-contamination of good news and bad.

It might be thought good, for example, that between 2000/1

and 2001/2 the proportion of municipal waste travelling the caveman route to holes in the ground actually decreased, from 79 per cent to 77 per cent, and that the proportion being composted or recycled increased from 12.3 per cent to 13.6 per cent. The reality, however, was that the graph still slanted the wrong way. The percentage might have gone down, but the total sent to landfill actually went up, from 22 to 22.3 million tonnes. At the same time, the apparently meteoric rise in the recycling rate – a near doubling since 1996/7 – still left the UK close to the bottom of the European league. By 2001, Switzerland had hit 52 per cent, with the Dutch, Austrians and Germans not far behind. If you looked at organic waste alone, Britain looked even worse – just 5.7 per cent recycled in 2002, compared with Austria's 75 per cent. Even the better-looking paper recycling rate, 47 per cent, looked feeble against Germany's 90 per cent.

You would have had more luck prospecting for gold on the Essex marshes than in looking for someone outside Defra with a good word to say about British waste policies. The government can, and does, point to improvements. In the following year, 2002/3, the proportion of municipal waste going to landfill continued to decline, from 77 per cent to 75 per cent. More significantly, if only by a shaving, so at last did the total amount landfilled – down from 22.3 million tonnes to 22.0 (back to what it had been in 1999/2000). The recycling rate crawled upward, too, from 22.4 per cent to 24.8 per cent. Good news, then. You couldn't say the graph had moved far, but at least it had turned a corner. And yet few people cheered. Three quarters of the nation's garbage was still taking the worst environmental route. More troublingly still, bearing in mind that the basic principle of European and UK policy is waste *reduction*, the bottom line – the total amount of garbage collected – had gone on rising, from 28.8 million tonnes in 2001/2 to 29.3 million in 2002/3, an increase of 1.8 per cent. Our response to the problem might have got a little better, but the problem itself was getting worse.

Politically, of course, it is no great vote-winner – no leader will ever ride to Downing Street on a cry of 'Garbage! Garbage! Garbage!' Unlike New Labour, the Conservatives and Liberal Democrats included recycling in their 2001 election manifestos, and look where it got them. No leader, however, wants to preside over a catastrophe. Given a hole-in-the-ground regime that historically looks less like waste disposal than waste accumulation, the moment of crisis looms ever nearer. We may still be some way from a consensus on what needs to be done, but at least no one now believes the issue can be ignored. Pressure for reform comes not just from the usual suspects – the European Commission, Greenpeace, Friends of the Earth – but most insistently from within Westminster itself.

Early in 2001, the Parliamentary Select Committee on Environment, Transport and Rural Affairs lifted the national dustbin lid and took a long, deep sniff at the government's policy. The priorities by now ought to have been clear. In drafting its policy, ringingly entitled 'Waste Strategy 2000', the government had protested its good intentions. The minister responsible, Michael Meacher, signed a foreword that, political grandstanding apart, might have been drafted for him by Friends of the Earth:

Ensuring a better quality of life now and for generations to come – 'sustainable development' – is at the heart of our programme. The economy is growing steadily, employment is at record levels, and we are making good progress on social issues such as rough sleeping. In March we announced that we are on track to meet and exceed our internationally agreed climate change targets. But if we are to deliver sustainable development it is crucial that we begin to tackle our growing mountain of waste. This means designing products which use fewer materials and using processes that produce less waste. It means putting waste to good use, through re-using items, re-cycling, composting, and using waste as a fuel. And it means choosing products made from recycled materials.

The policy itself was based on the 'waste hierarchy' set out in the EC Framework Directive on Waste – a penny-plain piece of common sense that placed the various strategies in order of environmental friendliness, from best to worst. At the top came waste minimization, espousing the truism that the best way to deal with rubbish is to produce less of it. Next best was reuse, followed by recycling and – significantly less good but better than nothing – burning rubbish to generate energy. Right at the bottom came incineration without 'energy recovery'; then landfill. What did this mean in practice? From ministers and officials, the environment select committee heard there was, indeed, to be a strategic change. Instead of the most environmentally damaging option, holes in the ground, the government planned to switch to the next worst – inciner-ation. The committee struggled to believe its ears. Having listened to expert witnesses describe garbage incinerators as carcinogen factories with scant control over their own output, it found itself almost at a loss for words. 'It is difficult,' it said, 'to fully express our disappointment . . . The majority of those involved with waste in this country appear to be guilty of thinking without imagination and planning without ambition.'

The national waste strategy, it concluded, was no strategy at all – just 'a list of aspirations' which the government was doing little or nothing to achieve. It complained that the government did not appear to take seriously the need to reduce waste, and that its recycling targets were 'depressingly un- ambitious'. As was its way, the government did not find it possible to accept criticism but – in an almost parodic exaggeration of New Labour management-speak – it did agree to 'review the delivery mechanisms'.

Two years later, in May 2003, the select committee – now known as Environment, Food and Rural Affairs – reported again on what it hoped would be the government's progress. Again it reeled away with a handkerchief clamped to its nose. It saw little sign of Whitehall being able to deal with the looming crisis, and little evidence of willpower. In particular, it deplored the 'glacially slow rate of change' and was appalled by the apparently pilotless drift towards incineration:

... we are still concerned that Defra's and the Environment Agency's lack of funds and expertise will delay real progress.

We are concerned that Defra still appears to lack the capacity, the vision, the sense of urgency and the political will to break the mould and bring about truly sustainable waste management in this country.

The Government should publish a report on the use of incineration techniques setting out the case both for and against this type of waste disposal. It should also make its own position clear on incineration, addressing particularly the health and environmental implications of this type of disposal.

'Waste Strategy 2000', for all its poverty of ambition, did represent a breakthrough of sorts. Earlier strategies in the 1990s had acknowledged the need for Britain to clean up its dirty-man-of-Europe act. They had set recycling targets but ensured their failure by not giving them statutory force. Whatever powers were needed to make us think more creatively, they were not the powers of prayer or wishful thinking. We needed legislation, enforcement, penalties. Unbelievably (or perhaps all too predictably), 'Waste Strategy 2000', too, was drafted without statutory targets. Only after pressure from campaigners was the policy given a legal backbone, though in comparison with the rest of Europe its targets still looked half-hearted. We were to recycle or compost at least 25 per cent of household waste by 2005, 30 per cent by 2010, 33 per cent by 2015.

And even now we were bigger on piety than pragmatism. You could set statutory targets for anything you liked – hours of sunshine, say – but it still left open the question of how they would be achieved. How could hope be transformed into reality? Would the secretary of state thread daisies in her hair and dance with her civil servants around a maypole on College Green? Might she offer a sacrifice (Michael Meacher, perhaps) to St Jude, patron saint of lost causes? Or would she just order some bigger wheelie bins? The environment committee wailed; the waste industry rolled its eyes and Friends of the Earth doggedly went on lobbying. It was rewarded in 2002 when the Labour MP Joan Ruddock introduced the private member's bill which, with government support, would become the Household Waste Recycling Act. By no later than 2010, it said, every local authority would have to collect from each household at least two separate recyclable materials. FoE didn't get all it was looking for (it wanted four or five different materials recycled) but at least it was a step in the right direction – away from landfill, away from incineration, and towards the far green hills of conservation.

Opposition to landfill is not all driven by the issue of dwindling resources. Even if limitless space remained for local authorities and contractors to dump their waste in, there would still be grounds for objection. For years, scientists monitored an apparent link between the most heavily polluted dumps – those containing hazardous industrial waste – and birth defects such as spina bifida, cleft palate, low birthweight and premature births. Then in August 2001 the *British Medical Journal* published findings from Imperial College, London, that raised doubts about *all* landfill sites, of which there were some 2,300. Babies born within 2km of such a site, it said, were statistically more likely to suffer abnormalities than those born further away. As the risk zones included 80 per cent of the population, this did little to offset landfill's tarnished image or to win political support. The risks may be small, and the statistical association may fall short of proof, but in the age of BSE, CJD and FMD the public had no appetite for bland assertions of safety. Neither is it the case that the links with disease are scientifically implausible. The metallic element cadmium, for example, has a known association with cancers of the lung, throat and prostate, kidney damage and kidney disease – and 10 per cent of the UK's entire emissions of cadmium come from landfill sites. There are wider issues too.

The decomposition of garbage produces waste products of its own – not just bad smells but 27 per cent of the UK's total output of the greenhouse gas methane, which is twenty times more powerful than carbon dioxide. Property prices are the most sensitive barometer. Early in 2003, Defra reported that houses within half a mile of a landfill site fell in value by an average of £1,600, and those within a quarter of a mile by £5,500. Nationally, it calculated, the total mean reduction in house prices within half a mile of landfills amounted to £2,483m – equivalent to between £1.52 and £2.18 for every tonne of landfilled waste. On every pair of lips from estate agents to Greenpeace, landfill of organic waste had become the ultimate no-no.

And yet . . . the total amount of municipal waste going for burial is as much as it was five years ago. For an almost unmentionable, politically unthinkable crime against the environment, landfill wears its stigma lightly. In the muck-and-brass culture of the garbage trade, its grip remains as tight as a scrap merchant's fist. And yet, come what may, by one means or another, the fingers will have to be prised open. The EC Landfill Directive requires the UK by 2010 to reduce the amount of biodegradable municipal waste sent to landfill to not more than 75 per cent of the total for 1995, with further reductions – to 50 per cent and 35 per cent – in 2013 and 2020. The government made a first attempt in 1996, when it did what it always does when it wants to discourage something. It imposed a tax. The result was predictable too. There was a Treasury windfall, a few unlooked-for side-effects, and not much effect on the target. The Landfill Tax, costumed in green, was celebrated as 'the first environmental tax introduced in the UK'. Anyone dumping in a licensed landfill site thereafter would be taxed at £7 per tonne for biodegradable waste and £2 for unbiodegradable. The higher rate went up in 1999 to £10, and was set to inflate by £1 per tonne each year until 2004, when it would reach £15 and the whole system would be reviewed. Almost no one believed this kind of money would be enough to deter the commercial operators, and so it proved – encouraged by the lowest landfill cost in any European country except Greece, the dumping continued to multiply. The industry's view was that the rate would have to at least double before the tax would bite, and the environment committee itself recommended £25 per tonne – a figure it would subsequently raise to £35. When the review came in 2004, the chancellor announced that the rate of increase for biodegradable waste would accelerate to a minimum of £3 a year from 1 April 2005. It brought the £35 target a little nearer – 2011 instead of 2022 – but hardly amounted to fiery breath on the dumpers' necks. This was despite a clear warning from the industry itself that, without rapid escalation to a higher rate, there would be no incentive for it to invest significantly in alternative technologies. Even a report commissioned for the Prime Minister's Strategy Unit in 2002 came to the same conclusion: 'Only a pronounced step-change in Landfill Tax could sufficiently close the cost gap with other existing and emerging added value technologies.'

Nevertheless, it would be wrong to imply that the tax had no effect. Sixty per cent of local authorities surveyed by the Tidy Britain Group reported an immediate increase in fly-tipping, often by more than 50 per cent. One South London borough recorded 4,256 incidents in a single year. Even Pendle, in rural east Lancashire, had 1,837. It goes on still, with the illicit trade expanding in parallel with the licit, tracking it like an albatross. Some of the dumping is in back streets and alleys, but most is on or near farmland. Most often the farmers are victims (they have to pay for the junk to be removed – an average of £1,000 a time); sometimes they are accomplices. With some farm incomes having sunk below subsistence level, and others on the brink, the temptation to diversify illegally into the dumping trade can be hard to resist. At one farm in Hertfordshire, the tenant allowed 30,000 tonnes of household, hospital and building waste to pile to a height of forty feet. The Environment Agency calculates that it shares with local authorities and landowners an annual bill of between £100m and £150m for clearing up after an average of 50,000 fly-tipping incidents. In criminal terms this is a great leveller – the one kind of environmental insult that allows the amateur polluter to compete on equal terms with the professionals. In order of popularity, the Environment Agency says, the ten most common categories of fly-tipped waste are bin-bags of household rubbish, garden waste, timber and fencing, builders' rubble, window frames, tyres, asbestos, Calor gas bottles, retail waste, and carpets, beds and bedding. In 1997/8, ENCAMS calculated that councils in England and Wales collected on average 556 cubic metres of illegally dumped household waste – equivalent to about 2,500 wheelie bins each. More than 70 per cent of local authorities say they have a 'significant problem' with fly-tipping, though only 23 per cent of them ever prosecute offenders.

In serious cases where it can trace the culprits (about 200 cases a year), the Environment Agency itself will take them to court. In March 2004, for example, a Swindon haulier was fined £15,000 for dumping three lorryloads of excavation and demolition waste – some sixty tonnes in total – on the historic but much-abused Ridgeway. What is especially depressing about this and other cases like it is the sheer nihilism implicit in its calculation – that an ancient landscape is worth no more than the £48 the man would have saved in tipping fees. More depressing still is that only a small minority of fly-tippers ever face the magistrates. The odd one will dump something that can be traced back to him, or will be turned in by witnesses calling the police or the Environment Agency's emergency hotline (0800 80 70 60). Most often, however, they chuck and chance it, and luck turns a blind eye to justice. In the same month, the agency appealed for help in finding the people responsible for a rash of offences near the Scammonden Water reservoir in West Yorkshire. There was nothing particularly unusual about what went on here – it happens daily throughout the country. And this is the real sadness of it. It is typical, not exceptional, an all-too-ordinary instance of one of the country's most visible but least remarked-upon criminal enterprises. On a single day in January, 500 tyres and a number of barrels containing mixed waste and car parts were dumped at the reservoir near Halifax. Two months later another 850 tyres appeared alongside the nearby A640. In each case the tyres were spread over a long stretch of road, suggesting that they had been thrown from moving vehicles. Why? 'What we suspect,' said the local environment officer, 'is that someone is running a tyre collection service. They are probably charging for the service and then, instead of disposing of the tyres by a legitimate route, they are dumping them at Scammonden.' Private profit: public cost. Cleaning the mess up cost Kirklees Council £5,000. Nationally, the public cost of tyre-dumping is running at around £2m a year.

It is yet another case of what might be called Edward III Syndrome, when a tax or prohibition confounds the law-makers' intentions. Inflating the price of riverside waste-tipping in the 1350s increased the rate of illegal dumping in the Thames. Putting up the cost of tyre disposal six-and-a-half centuries later invites a further application of the same unsophisticated logic. Landfilling whole used tyres has been banned since 2003, and the ban will be extended to include shredded tyres in 2006. Various ways are being tried to prevent the accumulation of a rubber mountain, but there is a petty-criminal, muck-and-money subculture that traces its antecedents via the horse trade all the way back to the medieval midden. For these boys, the commission of minor, largely undetectable and (as they would see it) victimless offences is of no more account than letting your rottweiler foul the pavement. Waste directives for such people are as remote as the *Tractatus* of Wittgenstein. You got some tyres you want to lose? No problem!

Defra in the meantime tries hard to sustain its optimism. 'The Government,' it says, 'appreciates the enormous impact the Landfill Directive will have on the disposal of used tyres but is confident that the extra tyre recovery capacity that will be needed as a result of the directive will be made available.' The world waits with interest to see how the cowboys will respond to the government's call for 'flexible and innovative responses'.

For the local authorities, it's a bit like trying to tidy an earth-quake with a dustpan and brush. To find and convict a fly-tipper you'd need all the resources of a murder inquiry – detectives, forensics, paid informers, media coverage – all adding up to the most radical rearrangement of the criminal landscape since Moses received the Ten Commandments. The Environment Agency's answer has been to take a leaf out of the urban policing book and go for the pre-emptive strike. In Birmingham, West Bromwich and Maidstone, for example, it joined forces with police and local authorities in a series of stop-and-search operations targeted at vans, pick-ups and trucks carrying anything that resembled junk – builders' rubble, old refrigerators, tyres, bits of metal, anything at all that looked no longer fit for its designated purpose. Commercial waste disposal operators need to be legally registered – a formality with which cowboys operating fly-tipping scams seldom see the need to bother. A simple request to see papers, therefore, is all it takes to net an offender. In a single day in Birmingham, fifty of the seventy vehicles stopped were not legally registered. It was a good day for the police, too. Fly-tippers are nothing if not consistent in their contempt for the law. They drive while disqualified, use stolen vehicles, fill their tanks illegally with agricultural diesel, drive brakeless, overloaded and untaxed lorries and carry hazardous loads. Officers soon had writers' cramp. After two days of this, the Environment Agency received forty 'unsolicited' applications for waste carriers' licences. It looks like a good result but, again, we must not forget Edward III Syndrome or the criminal mindset. Will the simple act of registration turn black hats to white? A fly-tipper with his papers apparently in order might slip through the cordon as easily as a corporation dustcart.

Usually, the celebrations after successful investigations are muted. The maximum fly-tipping fine in a magistrates' court is £20,000 but – to the Environment Agency's perpetual irritation – the actual penalties seldom rise above a tenth of that. Neither do the courts make much use of their other legal powers, to send offenders to prison or impose community service orders. The government meanwhile is growling in its corner and threatening a show of teeth. In early 2004 the 'local environmental quality minister', Alun Michael, announced that his department would be considering 'a range of tools' including fixed penalty notices for unregistered waste carriers, fines of up to £50,000 for repeat offenders and extra powers for local authorities to clear waste from private land and shut down illegal dumps. The Environment Agency and local authorities at the same time have beefed up their defences with a new national fly-tipping database, called Flycapture, designed to monitor trends and identify hotspots. It will need plenty of storage capacity. In July 2004, regulations came into force that threatened to put even greater temptation in the way of the casual law-breaker. The definition of 'hazardous waste' (i.e., stuff not allowed in wheelie bins) was extended to include creosote and paints. Warning bells rang at the Institution of Civil Engineers (ICE), whose annual State of the Nation report identified the obvious risk: 'Garden fences and old tins of paint become a problem and a cost to dispose of. Garden sheds, workshops, cellars and garages up and down the country will harbour a vast amount of this newly defined hazardous material. Fly-tipping must not become the easy option to dispose of it.'

A whole new, counterfeit world has been built on foundations of smuggled trash. All that crime needs to flourish is a market. If a company has a few tonnes of junk it wants to get rid of, and if someone else with a bit of waste ground or a half-built golf course is charging less for landfill than the official sites . . . well, why not? Peter Jones, a director of Biffa Waste Services, reckoned in 2002 that at least 3m tonnes of material had been illegally diverted to mould the contours of golf courses, shopping malls, sports complexes and housing estates. Many of the rogue dumps were actually registered with the Environment Agency as so-called 'exempt' sites, where it is legal to dump limited amounts of soil or builders' rubble for infilling or landscaping. As such sites are neither licensed nor inspected, what actually goes into them is any-body's guess. 'We are aware,' said the agency in 2001, 'of cases where illegal operators have opened sites alleged to be construction sites and in fact used them to dump contracted waste.' But it appeared to be fighting uphill against an over-whelming force of numbers. 'Active policing on the ground would take enormous resources,' it said. By 2004 it had some 65,000 such sites still on its register, but could not say how many of them remained active. At last, however, there was some chance of improvement. In the late summer of 2004, new regulations came into force that gave owners or operators of exempt sites six months to re-register them, with each registration in future lasting for one year only. Any site not re-registered will have its registration cancelled, and there will be tougher penalties for rogues. The intention is to achieve a double benefit – tighter control, and an end to the bureaucratic nightmare of a constantly escalating number of registered sites. 'The effect,' says the agency, 'is that there will be auto-matic removal of a registration if there is no renewal notification, and therefore the register will be up to date.'

Well . . . ho hum, maybe, let's hope. One thing the agency does not have the resources to do is inspect 65,000 formerly exempt sites. To imagine that the men who worked these scams would mothball their skips just because their registration has expired is to show a faith in human goodness that would test the credibility of a lapdog. What have they got to lose? Like fly-tippers, garbage buccaneers are seldom brought to court. Even when they are, the penalties are not much of a frightener to men who are generally of the thicker-skinned type, and whose ticket through life is a wad of used notes. Fines typically are around £1,000, which, says Biffa's Peter Jones, 'means nothing'. And it's not just the criminals who have profited. Even legitimate waste disposal companies have been engaged in an ingenious wheeze operated in full public view by the government. The extraordinary – some would say farcical – Landfill Tax Credit Scheme was supposed to promote 'green' alternatives to landfill. Hole-in-the-ground operators were allowed to claim hefty tax rebates in return for supporting environmental trusts that spent the windfall on 'green' improvements. The fatal flaw was that the government decided to regard the rebates as private rather than public money, and allowed the operators themselves to choose which schemes to support.

Unsurprisingly, the geese did not vote for Christmas. Landfill operators are in business for profit, not kudos. They could hardly be expected to favour the kinds of enterprise – recycling schemes, say – that might knock the enamel off their own core businesses. Given a choice, they much preferred projects that were compatible with their landfill operations, and the best way to make sure this happened was to set up environmental trusts of their own. In the spring of 2000, the *Guardian* revealed a catalogue of abuses that ranged all the way from deliberate fraud by contractors to mismanagement by the credit scheme's private regulator, the optimistically named Entrust, whose board at the time was dominated by the waste disposal industry. Local authorities and waste contractors were partners in a mutually beneficial merry-go-round of bribery and coercion. A contractor would say to a council: give us planning permission for a landfill extension or we'll spend our tax credits somewhere else. A council would say to a contractor: spend your tax credits here or you'll lose the municipal waste disposal contract. Tax credits were used for, among other things, a new access road to a landfill site in Scotland, and funding anti-recycling propaganda for landfill and incineration businesses.

In March 2002, Entrust's chairman and chief executive were charged with contempt of parliament and had to make a public apology. Their offence had been to offer 'threats and molestation' to a witness. An environmentalist summoned by the environment select committee to give evidence had been dismissed by Entrust for suggesting that the tax credit scheme was not being properly run. This happened *after* the committee had censured Entrust for 'immature and unprofessional' behaviour, and 'entirely inappropriate sensitivity to criticism'. The MPs called for it to be replaced by a more representative public body, and for an end to the 'charade' of the Landfill Tax Credit Scheme – demands which the government met with neither the hoped-for U-turn nor, quite, the expected deaf ear. The Entrust board now includes no one recruited from the waste disposal industry. A new chairman and chief executive are in place, and the Landfill Tax Credit Scheme has been reformed, with £110m a year being siphoned off to help pay for the government's new 'sustainable waste delivery programme'.

*

Landfill's heir apparent, incineration, meanwhile was making news of its own. After smoke, ash and dioxins, its principal by-products were fear and confusion. Ministers, civil servants and industry experts giving evidence to the select committee in 2001 could not agree even on how many municipal incinerators would be needed in addition to the twelve already burning. Was it 166 (the government's own first stab)? Was it 140? 112? 100? Four? Whose backyard should they be in? How big should they be? Were some materials simply too dangerous to burn? If so, what should happen to them? And what about the ash? Given recent disasters, and warnings that incinerator ash might be ten times more carcinogenic than Agent Orange, how could the public ever be persuaded to accept them? Apathy vanished like spit from an Aga whenever they were mentioned. All over the country, encouraged in particular by Greenpeace, local action groups were digging in behind the legal barricades: Guildford, Kidderminster, Richborough, Portsmouth, Newhaven, Hastings, Bexley, Havering, Swale, Beddingham, Whiteparish, Nottingham, Colchester, Chelmsford, Rayleigh, Basildon, Hull, Stallingborough, Wrexham, Port Talbot . . . On and on, a nationwide gazetteer of dismay.

Facing outrage in so many constituencies across the country, the government shied away from its own forecast. It began to deny that there was any target at all, and said that in any case it was not Whitehall's responsibility – it all depended on local authorities, and how good they were at recycling. When the government's own statistics were disentangled, however, it became clear that they expected as much as a third of the country's rubbish to be going to incinerators by 2010.

This is burning on a colossal scale. Mass-burn municipal

incinerators are huge industrial plants, the size of power stations, looming over their neighbourhoods like satanic throwbacks to the nineteenth century. In terms of air pollution the new ones are significantly less damaging than the old, though reassurances of good neighbourliness from local authorities, government and waste contractors do not carry a high plausibility rating. Public distrust has been well earned. The story of municipal incinerators in the UK has been a black comedy of cowboy capers, GBH and regulatory farce, marking the locations of some of the worst pollution blackspots in the country. In Newcastle, for example, the worst address was anywhere within sight or scent of the huge incinerator at Byker – a name that hangs over the waste industry like a shaman's curse. If Blair and Bush had found anything like it in the Iraqi desert they would have declared their policy on WMD fully vindicated. The only comfort may be that events in Newcastle were so apocalyptically dreadful that nothing like them could be allowed to happen again, anywhere, ever. Well, maybe . . .

The real problem with incineration is that it produces ash – mountains of it – and it is not harmless stuff like you get from a domestic grate. Household waste is laced with horrors – insecticides, pesticides, cleaning fluids, bleach, dyestuffs, batteries, fluorescent light tubes, televisions, computer screens, keyboards, lead solder, wood treatments, PVC . . . Burning them releases heavy metals and toxic chemicals that collect and become concentrated in the ash. The very worst of them, the so-called dioxins created by burning chlorine (an ingredient of household cleaners and PVC), are nightmarish carcinogens whose uncontrolled release into the environment is akin to chemical warfare. For this reason, contaminated ash is usually sent to specially designated landfills. Newcastle

City Council, however, believed it had a better idea. Between 1993 and 1996 it dumped more than 2,000 tonnes of Byker's toxic ash on to footpaths, parks and allotments. Analysis revealed dioxin levels 2,000 times higher than the recommended safe limit for topsoil. Levels in eggs from allotment hens were high enough to double the cancer risk of anyone who regularly ate them. The city council and the company that operated the incinerator, Contract Heat and Power Ltd, admitted offences against the Environmental Protection Act, and after many delays were fined a total of £30,000 with £35,000 costs at Newcastle Crown Court in January 2002. The plant now burns only what the council describes as 'conventional fuels' to generate power for local housing; waste is no longer sent there.

'A national disaster' was how one expert described Byker. Yet surely this was an isolated act of madness, a one-off with no possibility of recurrence? Only a 'complete berk', argued Peter Jones, would have continued to mix the highly toxic 'fly ash' from incinerator chimneys with the relatively safe 'bottom ash' from the grates. Well, maybe . . . But then came Edmonton, in north London. The giant incinerator here is run by LondonWaste, a company formed by the French-based company SITA with seven local borough councils. Its wheeze was at least a bit more sophisticated than Byker's. Instead of broadcasting the ash as a cheap 'soil-improver', Edmonton sent it to be made into roads, car parks, bricks and breeze blocks – a chemical booby trap awaiting any builder, navvy or do-it-yourselfer with a drill or hammer in his hand.

Nobody knows for certain where all the stuff went; whose roads are paved and whose houses are walled with dioxins. Grilled by Jeremy Paxman on the BBC's *Newsnight* programme, the environment minister Michael Meacher displayed all the symptoms of shock and said he would take action 'first thing tomorrow'. He promised an urgent report on where the ash had gone, and did not rule out the possibility of prosecutions. That was in July 2001. It was not until May 2002 that the Environment Agency published a sixty-five-page report accounting for all the ash produced over a five-year period, 1996–2000, by all eleven municipal incinerators. It found that Edmonton was the only 'complete berk' to have mixed fly and bottom ash, and that, before it ended the practice in 2000, some 15,000 tonnes of it had gone into an estimated 5,337,500 construction blocks – enough for 3,400 houses. The blocks themselves were by now untraceable (the material had been sold to seven different companies in places as diverse as Arundel, Doncaster, Teeside and Thirsk) but the agency saw a need only for sensible precautions, not panic. The blocks would not release measurable amounts of dioxins into the rooms that were built of them, and do-it-yourselfers using drills would be OK so long as they took the 'usual precautions, such as wearing suitable respiratory protection, using a dust-sheet and cleaning up carefully afterwards'. Beyond that, there was not a lot the agency could do. It had a responsibility for policing waste, but ash used in building materials ceases to be waste and becomes a 'product', thus escaping the agency's jurisdiction. Comment in the circumstances seems superfluous, though the report's authors did not flinch from stating the obvious: 'Government should consider the need for guidance on acceptable contaminant levels in construction materials in terms of potential impacts on health and the environment.'

It is true that incinerators have improved. The new ones burn more cleanly than the dinosaurs of old – most of which were closed down when the EU set new emission standards in 1995 – and the levels of dioxins, heavy metals, dust particles and acid gases pluming into the atmosphere have all been reduced. Nevertheless, truths in the waste business are seldom absolute. Even the reduced levels are arbitrary. The legal limits do not meet any standards set by a health authority, for no such standards exist. Rather, they represent the best that current technology can achieve – precisely the same basis on which walking a tightrope was once the best route across Niagara Falls. There is no consensus on what is safe. In an effort to metaphorically if not literally clear the air, Defra commissioned the environmental consultancy Enviros and the University of Birmingham to review the evidence – not just on incineration but on the environmental and health impacts of all existing waste disposal systems. Their report was peer-reviewed by the Royal Society and published in May 2004.

It soon confronted the age-old problem: 'There is no lack of research reports and reviews, although it can be hard to find information of good enough quality to rely on.' Among those whose work it examined were the Environment Agency, the Irish government, the American Agency for Toxic Substances and Disease Registry, Greenpeace and the UK government itself. Its conclusions were reassuring. Although it conceded that the older, more polluting incinerators may have had a bad health record, it found no such link with the new. 'We considered cancers, respiratory diseases and birth defects, but found no evidence of a link between the incidence of disease and the current generation of incinerators.' There is no reason to doubt this, though it may yet be too soon to sleep with the window open. Cancers do not appear overnight. The old incinerators had decades to sow the seeds of premature death, and it may be ten or twenty years before the new ones have earned the right to be regarded, however grudgingly, as good neighbours. The report's authors also seemed unconcerned about mercury (of which incinerators contribute 20 per cent of local background levels) and dioxins. While they made no attempt to downplay the toxicity of dioxins – as well as being carcinogenic, they are known to arrest the development of male reproductive organs, cause other birth and developmental abnormalities and damage the immune system – they argued that the amounts discharged by modern incinerators are not significant. 'We found that dealing with municipal solid waste accounts for only about 1 per cent of UK emissions of dioxins, shared approximately equally between incineration and emissions from burning landfill gas.'

Against this, they said, 18 per cent of UK dioxin emissions came from coal-burning domestic grates and cooking. The logic of this is curious. It seems to suggest that polluting the air is excusable if you're not the worst offender, and that straws never break camels' backs (though the report did rank dioxin pollution from incinerators as 'moderate' rather than 'good'). Nothing will hang in the air longer than the scepticism borne of the industry's past record. Historically, the operators' attitude to emission controls has fallen somewhere between nonchalance and contempt. In 1999 and 2000, for example, the eleven municipal incinerators in England between them infringed their limits 543 times. Yet again in September 2001 the Environment Agency had to serve enforcement notices at Wolverhampton, Coventry and Edmonton. In the first nine months of 2004 there were thirty-four breaches, of which fourteen required formal action from the Environmental Agency. Despite Defra's best efforts, it remains difficult to know how bothered we should be. Health research, as so often, yields more contradictions than conclusions. One report for the European Commission made the startling claim that every tonne of municipal waste burnt in an incinerator cost between £21 and £126 in damage to public health and the environment. Yet it is notoriously hard to nail down cause and effect. Some surveys appear to show that children living near incinerators have suffered more respiratory disease, allergies and colds; others find no difference. The US Environmental Protection Agency warned that dioxins might be 1,000 times more toxic than anyone previously had thought. Increased cancer risks have been reported from areas near incinerators in Italy and France. In the UK, increased risk of laryngeal cancer was associated with one incinerator where hazardous waste was burned but not with nine others. Other UK studies found a 37 per cent increase in liver cancer, a doubling of childhood cancers, an increase in the rate of multiple births, an increased ratio of female births, and increases in cleft lip and palate, spina bifida and other congenital defects in children. Others found no effects at all.

Every attempt at clarification serves only to deepen the confusion. Even the government's hired consultants, Entec, added to the chaos by suggesting that every million tonnes of incinerated waste would result in twenty premature deaths and forty-one hospital admissions – figures which, grossed up nationwide and with a potential 9m tonnes in the system every year, looked like a mass slaughter policy. Fortunately this was a mistake. Once they had put a new battery in their calculator, Entec had to admit that they had blundered by what the environment select committee in 2001 drily called 'almost an order of magnitude'. The revised version was that fewer than three of the 24,000 premature deaths attributed every year to air pollution could be blamed on incinerators.

So who and what can we believe? As we have seen, all incinerators are not alike. Even in the past, some burned more cleanly than others, which is a plausible explanation for at least some of the discrepancies in the research. Variable pollution, variable risk. And there are plenty of statistical 'confounders' – other possible explanations for observed effects. Incinerators tend to be in poorer parts of town where people are less able to defend themselves than the better-organized and more powerful professional classes. As poverty is the biggest risk factor in almost every disease you care to mention, you cannot discount the possibility that people in these places – having the worst diets, the worst education, the worst nicotine habit and the worst medical care – would have poor health even without an incinerator. Toxicity, however, is not the only issue.

Incinerators are offensive to every fibre of a conservationist's being. They are aesthetically hideous, no less degrading to their neighbourhoods than nineteenth-century dust-yards. They are unsustainably wasteful – into the searing maw goes tonne after tonne of material, much of it derived from fossil fuels, that could be reused, recycled or composted. They are material insults to the long-term stability of the planet. Not only do they contribute directly to global warming but they encourage the over-consumption of fossil fuels in the creation of yet more waste. Both Greenpeace and Friends of the Earth campaign for a zero-incineration economy, and Greenpeace in particular has made a speciality of direct action, occupying municipal incinerators and forcing them, if only temporarily, to damp the fires. Intellectually if not physically, they keep good company. While the hon. members of the Select Committee on Environment, Food and Rural Affairs are unlikely to shin up chimneys with 'Cancer Factory' banners, their report in 2003 contains a reverberant echo of the activists' case.

The seamless robe of joined-up government is as mythical as the holy grail it is often likened to. There always have been, and always will be, policy conflicts between departments – most often between the Treasury, which wants to save money, and others that want to spend it. Short-term savings are often at the expense of long-term efficiency. Usually, however, there is some underlying thread of consistency and a vague notion of where we are headed. Everyone can agree on the destination, even if they follow divergent routes for different reasons at different speeds. Waste policy, however, has been less a drifting vessel than a rudderless flotilla whose members are forever colliding with each other. Whichever way you turn, you meet someone else coming the other way.

Defra insists it is fully committed to the big leap in re-cycling rates that Europe has demanded. To succeed, however, it will need a balanced framework of penalties and incentives to push industry, local authorities and the rest of us into chang-ing our ways. Being only human, we'll go on chucking our muck over the fence until someone makes us stop. Without the right balance of encouragement and coercion, waste minimization will join hands with veganism, homeopathy and nettle soup in the sunshine province of the sandalistas. The problem is that all the pressures currently work in the opposite direction.

'At present in England,' said the environment committee, 'the price of waste management options is almost the reverse of the waste hierarchy; landfill is still by far the cheapest option, incineration is often the next cheapest and recycling is considered very expensive . . . We fear that increases in the landfill tax could simply drive waste to the next cheapest option, which is likely to be another form of disposal such as incineration. We would like to see a system where the relative net costs of waste management options reflect their position in the waste hierarchy.'

This raised a loud and sustained hurrah from within the industry itself. Biffa Waste Services issued a statement from its director Peter Jones. The committee's report, it said, was 'yet another independent indictment on the economic failure occurring within the sector. This country cannot move to more sustainable waste management unless the government addresses the availability of cheaper, less sustainable exit routes for waste. The intransigence of the Treasury to take this message on board is inexplicable.'

Jones, like the committee itself, is a persistent critic of official myopia. He went on:

Even when tax instruments create a level playing field – some time around 2010 on current forecasts – the demand for sustainable technologies threatens to overwhelm equipment suppliers . . . Thousands of new facilities will need to be commissioned over the next decade as landfills fill up – leaving it to the last minute will create chaos. Ironically, around £500m–£600m each year is being pumped in as state aid to subsidise the achievement of short-term targets – mainly on municipal waste. Yet there continues to be manifest resistance to the use of market instruments which would achieve those objectives far more effectively. As the targets progressively rise, how deep is Gordon Brown's subsidy pocket? The government is in danger of reinforcing the view yet again that it is incapable of strategic thought and direction. This is perverse given that the technology and a broad consensus exists between leading waste companies, non-government organisations and local authorities on the broad thrust of how to tackle the looming crisis.

Even this grim analysis may have been over-optimistic. The select committee complained that Defra had so comprehensively failed to develop and clarify its strategy that those given the responsibility for carrying it out – primarily local authorities – had little idea of what was expected of them. 'We do not yet know,' said the Local Government Association, 'what we are going to be asked to deliver . . . We are still not getting coherent, joined-up strategic thinking.' The Environmental Services Association complained that Defra simply did not understand the business, and that it had displayed an 'incredible lack of engagement'. The Organization for Economic Co-operation and Development (OECD) meanwhile reached the unavoidable conclusion that 'measures to encourage waste minimisation remain very weak'.

Part of the problem is that (except for Unitary Authorities, which do the whole job themselves) responsibility for the collection and disposal of refuse is split between different authorities. District councils collect; county councils dispose. Recycling is the responsibility of the collecting authorities, to which the disposal authorities must pay 'recycling credits' equivalent to the costs they save by not having to dispose of the material that has been recycled. The net effect of this is that county councils have zero incentive to encourage or promote recycling because their savings all have to be repaid to the districts. Worse: market and fiscal pressures pull in the wrong direction – to maximize waste and minimize recycling.

Incinerators are enormously expensive to build – typically around £80 to £100m – which means that local authorities have little or nothing left to invest in other strategies. The problem is exacerbated by the apparently 'green' policy of using heat from incineration, as at Byker, to produce hot water and electricity. This means that the contractors need a guaranteed supply of waste for burning over a twenty-five- or even thirty-year contract period, with compensation payable if the council defaults. Thus local authorities create for themselves a vested interest in *maintaining* the flow of garbage rather than reducing it, so that recycling rates are almost always lower in areas with incinerators. At Cleveland, for example, the county council agreed to supply its contractor each year with 180,000 tonnes of refuse for incineration and 80,000 tonnes for landfill. A 12,000-tonne shortfall in the first year cost the council a £147,000 penalty. Stockton Borough Council put it succinctly. 'Essentially,' it said, 'we are into waste maximisation.'

Also howling at the moon was the Department of Trade and Industry, by whose logic the incineration of biodegradable waste was classified as a source of 'renewable energy'. This not only exempted it from taxation under the Climate Change Levy – a tax designed to put the brakes on heavy energy users – but also qualified it for some £420m of additional public subsidy. Conveniently, it will also help the DTI to meet the government's target of generating 10 per cent of the nation's electricity from 'renewable resources' by 2010, a travesty that drove the select committee in 2001 almost to apoplexy.

'A waste stream,' it said, 'is sustainable only in the most twisted definition of the word, since sustainable waste management has as its cornerstone the minimisation of waste, and the explicit maintenance of waste streams for the purposes of incineration is in complete contradiction with this principle.' In 2003, in commendably restrained language, the committee said it again. 'The current economic and fiscal regime,' it complained, 'has failed to promote sustainable waste management . . . We recommend that the government ensure that all economic instruments – both taxes and subsidies – are used in such a way that they reflect the position of each waste management option in the waste hierarchy.'

There is much talk of levelling the playing field. Friends of the Earth argues that there is an inherent bias towards incineration in the way the systems are costed. The price of incineration, it says, is made to seem artificially low because, much like the price of intensively farmed food, it does not take account of the costs to health and the environment. The price of recycling at the same time is kept artificially high because it does not take account of *benefits* to health and the environment – minimizing the extraction of oil, minerals and other primary materials; reducing the need for processing, transport and manufacture; cutting pollution and reducing greenhouse gases. Even the energy saving from heat generated by incineration, it insists, is a false economy. 'The energy it produces is significantly less than the energy saved through recycling.'

Another objection to incineration is that it produces waste of its own. There is a misconception, implicit in glib talk of 'disposal', that if you burn something it disappears. But of course it doesn't. Between 15 per cent and 20 per cent of the stuff that goes into mixed-waste incinerators – rubble, metals, water – is non-inflammable. Even when it does burn it doesn't vanish but only alters its form. Modern incinerators may release fewer toxins into the air, but this doesn't mean the toxins themselves have ceased to exist. They are left behind, trapped in the ash. It was this, not air pollution, that drove the scandals at Byker and Edmonton. Every hundred tonnes of incinerated garbage leaves behind a third of its weight – thirty-three tonnes – in ash which, in a plant processing up to 600,000 tonnes a year, is on the scale of a minor volcano. As it is no longer possible to pretend that dioxins are suitable for spreading on allotments or building houses with, there is only one thing to be done with it. You need a hole in the ground.

How, then, do we drag ourselves out of the pit? We cannot reinvent ourselves; cannot purge ourselves of the instincts – greed, selfishness, imagination – that brought us down from the treetops and turned us into the most ingenious as well as the most damaging organisms in the known universe. From nutshell to clam-pack, waste has been an essential part of the consumer economy, in which the appetite is always for more and better. 'Sustainability' is an idea, not an instinct.

Sir Thomas More unintentionally bequeathed to reformers of all kinds an obstacle to credibility. The very word 'Utopia', deriving from the Greek for 'no place', is a millstone for anyone with a vision of perfection. It stands for the unattainable and exists only in the minds of holy fools and ideologues. It speaks Esperanto, wears what it spins and leaves not a stain on the earth. In Utopia there are no wheelie bins. Utopia is zero waste.

It does not help that environmental campaigners embroider the idea with capital letters – Zero Waste – in the manner of a religion. It is a state to which they not only aspire, but for which they have a timetable and a target – the year 2020, according to Greenpeace. What it means broadly is, stage by stage, designing waste out of the system – reusing and recycling more and more, disposing of less and less, until, at the moment of garbalogical nirvana, nothing at all remains to be burnt or buried. Is it achievable? In a sense, it doesn't matter. Every step towards paradise makes life a little better, and halfway is a lot better than no way at all. Greenpeace and Friends of the Earth, it hardly needs saying, are as strongly committed to Zero Waste as they are united in their opposition to incineration. It certainly sounds good:

Zero Waste is an extraordinary concept that can lead society, business and cities to innovate breakthroughs that can save the environment, lives and money. Through the lens of Zero Waste, an entirely new relationship between humans and systems is envisaged, the only one that can create more security and wellbeing for people while reducing dramatically our impact upon planet Earth. The excitement is on two levels: it provides a broad and far-reaching vision and yet it is practical and applicable today.

Zero waste is a concept that is spreading throughout the globe. Our vision is to reduce consumption of goods by ensuring that products are made to be reused, repaired or recycled. What we now call waste should instead be regarded as a mixture of resources to be used again to their full potential, not as something to be thrown away.

This is, you might think, typical production-line propaganda, the kind of stuff churned out every day by desk-dwelling idealists who have no responsibility for anything but their own, cost-free moral superiority. Airy-fairy, idealistic, blind. What a surprise, then, to find that both quotations come from government-endorsed websites and are statements of official policy. The first of them, it's true, is from New Zealand and not the UK; but the second is as English as toasted muffins – Bath and North East Somerset Council. In September 2001, Bath became the first British local authority to formally adopt a zero waste policy, and in 2002 New Zealand became the first national government to follow suit. Bath took another pioneering step, too, in dropping the capital letters, preferring the solid practicality of plain 'zero waste' to the theological – and, hence, implicitly theoretical – spin of the upper case.

You can argue about whether or not the policy is well named. 'Zero' is an absolute, and no regime on earth will ever achieve 100 per cent compliance from 100 per cent of its people. That is its weakness – the unreality of utopianism. Its strength is that it stands for a new way of thinking, requiring us to accept that waste is a shared problem and that shared problems need shared solutions. We're all in this together: it's not just a baton-pass to the waste disposal industry. The journey ahead, to the mythical zero, is less like a timed flight than a voyage under sail. Being subject to forces we cannot predict, we have a destination but not a time of arrival. In setting targets, in looking forward even to the next decade, we risk all the pitfalls of futurology. The greatest unknown is the impact of our own ingenuity. We cannot know how the pace of change will respond to the development of new materials, or of new uses for old ones, or of products and processes as remote from us now as cloned sheep and the silicon chip were for the generation that fought the Second World War. And this is the crux of it.

Garbage men have a technical term for the kind of policy we've followed ever since man first trekked to the midden – 'end of pipe'. It is a policy based on disposal. Whatever society discards, the bin-men will collect; and whatever the bin-men collect, the disposal industry will bury, burn or re-cycle. Zeroism looks at the issue from the other end. 'Reducing waste,' says the New Zealand Waste Strategy, 'cannot succeed without a system that manages waste from the point of generation through to disposal.' In other words, the solution begins with the first stroke of a product designer's doodle. It is no longer sufficient to guarantee that a product will do what it says on the packet. Manufacturers will have to accept end-of-life responsibility for their own creations, and be as involved in what comes out of the pipe as they are with what goes in. It means designing and choosing materials with maximum potential for future life, and leaving the smallest possible footprint on the surface of the planet. This is reality, not utopianism. It may tread lightly on the consciousness of the UK government, but it is already the basis of real European directives on the afterlife of batteries, electrical goods and cars (see Chapter 7).

In making their case for Zero Waste, the New Zealanders offer examples that might persuade even a Texan to hold on to his ring pull. Since 1990, for instance, the US has wasted more than seven million tons of aluminium cans – metal with a mar-ket value of seven billion dollars, sufficient to build 300,000 Boeing 737s. In manufacturing generally, 94 per cent of the materials extracted to make durable products become waste even before the products themselves leave the factory, and 80 per cent of what we make is junked within the first six months of life. Making a laptop computer, for example, creates nine tonnes of waste. But nothing illustrates the curse of convenience better than the disposable nappy. The plastic in each one costs a full cup of crude oil, and the pulp for the absorbent filler costs 4.5 trees per baby over the average two-and-a-half years of its nappy-wearing life. It takes as much energy to make one disposable as it does to wash a cloth nappy 200 times. And so on. It's not difficult to make the argument. The real problem, particularly when it involves effort, is to persuade people to face up to its logic.

The members of the UK parliament's environment select committee were keen to turn the fiscal screws – to jack up the landfill tax at a much faster rate than the £3 announced by the chancellor, and to adopt a differential tax system on the Danish model, in which each disposal system is penalized according to the damage it causes. Landfill in Denmark is in the highest tax band (with some materials banned from landfills altogether); incinerators pay a lower rate, but are required by law to generate heat and power. Recycling and composting are zero-rated.

It is a sad truth that 'best practice' is always represented by examples from other countries. No government anywhere in the world is looking to the UK for its model. The long backlog of bad policy-making, indecision and inertia means that significant waste reduction in the short term has already drifted out of the real world into fantasy. At the current rate of progress there is next to no chance that the 2010 EU target for reducing landfill will be met, and every chance that the UK thereafter will be punished for its failure by annual £180m fines. Some other countries – Denmark, Austria, the Netherlands – are so far ahead that they have already met the target for 2020. Of the sixteen existing member states before the expansion of the European Union in 2004, only Ireland, Greece and Portugal sent a higher proportion of municipal waste to landfill than the UK. At this late stage the need is less for the perfect waste policy, or even for a coherent, joined-up policy, than for any policy at all. The 'waste management' section of the Institution of Civil Engineers' 2004 State of the Nation report was a nine-page shriek of horror. It warned that the 'crunch point' of 2010 – the first landfill target year – was already well above the horizon, and that to meet the requirements of EU directives over the next sixteen years the UK would need between 1,500 and 2,300 new facilities to treat, recycle and dispose of its waste. As it would take at least five years to bring a single one of these on stream, its headline warning – 'For the Cinderella of the nation's infrastructure, time is running out' – could hardly be accused of hyperbole. It pleaded above all for rationalization. Policy needed to be re-engineered – straightened out and streamlined, not endlessly tinkered with and cluttered with bolt-on accessories. ICE itself counted no fewer than seventy-two 'separate and largely uncoordinated initiatives' funded either directly by government or by government-funded bodies. And heads needed to come out of the sand. John Prescott's planners were no more forward-looking with rubbish than they had been with water. 'The Office of the Deputy Prime Minister,' said ICE, 'does not monitor if enough waste treatment recycling and recovery infrastructure is receiving planning permission to meet the nation's needs. It is therefore extremely difficult to calculate the gap between the UK's current waste management capacity and that required to meet future needs.'

You can see why Prescott's people wanted to keep their heads down. In political terms it would be like riding with the Light Brigade – Cannon to right of them, / Cannon to left of them, / Cannon in front of them / Volley'd and thunder'd. The braver the policy, the heavier the shellfire. You can't solve a waste crisis without providing more facilities; you can't provide more facilities without gross offence to the electorates in whose backyards they are to be built. The Defra-commissioned Enviros/University of Birmingham report, 'Review of Environmental and Health Effects of Waste Management', had the effect of renewing local authorities' confidence in incineration. If this translates itself into policy, public opposition will be fierce, delays will lengthen and land-fill targets will be further away than ever. In the bleak lexicon of thwarted planners, NIMBY (Not in My Back Yard) has been supplanted by BANANA (Build Absolutely Nothing Anywhere Near Anyone). In November 2002, a report prepared for the Prime Minister's Strategy Unit by waste consultant Dr Stuart McLanaghan described incineration as 'politically undeliverable' (he blamed 'sensationalist media coverage'). This won't stop local authorities trying. Already, said ICE, major projects had been blocked or abandoned in Worcestershire, Norfolk, Hampshire and south-east London – though it seemed to see these less as victories for democracy than as defeats for reason. As an essential first step on the long road towards efficiency, it urged local authorities to earmark specific sites for waste handling, and deplored the example of London where, in 2003, the Greater London Authority reported that twenty-four of the thirty-three boroughs had failed to include them in their local plans.

With incineration so high on the public hate list, what to build is a bigger question than where to build it. Even a Zero Waste policy would have to provide for the processes of disposal to continue for as many years as it might take to reach the target (even the optimists are talking about 2020). While rates of reuse, recycling and composting may increase, we still have to cope with the 'residual waste' that's left behind. And as Zero Waste at the moment is about as far from the official mind as the planet Neptune, there are some very big questions jostling in the queue for answers. The Strategy Unit's prediction is that, despite all efforts to the contrary, the overall output of municipal waste will grow so rapidly over the next fifteen years that improvements in waste minimization, recycling and composting will struggle to keep pace. The bleak reality is that by 2020 the volume of 'residual waste' will be much the same as it is now. The best we can hope for, in other words, is that the problem will not get any worse. So what are we going to do with it all?

There is no easy answer. 'It is evident,' said the McLanaghan report, 'Delivering the Landfill Directive: The role of new and emerging technologies', 'that all technologies considered possess both environmental advantages and disadvantages. As such, there can be no Holy Grail solutions to managing society's waste.' Again, the official voice echoed the green lobby's. 'Achieving both more sustainable waste management and development will necessitate a more holistic approach to economic activity and the consumption of goods and services. This will require society to evolve economic activity to be analogous to the workings of natural ecosystems.' At this rate, they'll soon be invoking Gaia.

But how and when will any of this translate into action? Nearly two years after McLanaghan, in June 2004, Biffa's Peter Jones expressed the industry's mounting frustration:

> ... frustration attributable to the fact that despite the unique advantages possessed by this country – academic excellence, strong financial investment markets, high density of population – we are still trembling on the brink, racked by indecision, conflicting policies, imprecise signals and a poorly informed populace. The waste sector ... has the know-how, financial muscle and available tools to deliver the solution. So why is nothing happening?

One reason is the extreme caution of those 'strong investment markets' and the government's addiction to the smoke-and-mirrors illusion of PFI in preference to old-fashioned public funding. City fat cats choose their feeding bowls with care, and they like them deep and safe. The problem with new technologies is just that – they are new, with no proven track record of profitability. Add in all the other potential pitfalls – problems of siting, planning permissions and operating licences, plus uncertainty over the volume and nature of future waste – and you can see why banks and venture capitalists are not knocking each other down in the rush to invest. Incineration? Yup, that works, we'll put money into that. But advanced thermal treatments? Come back when you've got the numbers.

Yet again, humiliatingly for a country whose self-opinion not long ago rested on the boldness of its innovation and the excellence of its engineering, we have to find our exemplars abroad. A wide range of technologies exists – some of proven value, some still developing, others scarcely beyond the drawing board – but the one thing they have in common is that they have been pioneered in other countries. On its own, no single technology does the whole job. All leave some kind of solid residue to be disposed of, so that a total end to landfill is as remote a prospect as the total abolition of waste. Most involve some kind of heat treatment and the reclamation of heat or energy, sometimes in the form of gas. Many render biodegradable waste into a form that makes it suitable for composting. Some have the advantage of working well on a small, local scale, thus reducing garbage miles, serving the 'proximity principle' and minimizing their impact on the environment. Some have the disadvantage of requiring a guaranteed supply of combustible waste and thus weakening the incentive for recycling. For local authorities gazing into an uncertain future, the issues are far from simple. Legislation has not only to be complied with but also anticipated. What will be the effect of current and future EU directives? How much of the energy they generate will count as 'renewable'? When environmental benefit has to be balanced against capital cost, how can they meet their obligation to achieve best value? In a rapidly changing world, with new manufacturing and recycling technologies forever altering the garbalogical landscape, how should they calculate the balance between the costs of operating the plant, the fees to its clients and the income from its products? How can they simultaneously encourage home composting and operate a waste treatment system that relies on a steady supply of organic waste? How can they justify investing in systems that lack proven track

214

records? Wherever in the world you might expect to see the pioneering application of radically new technology, it is not in English local government. Methods now being talked about include:

composting – either in the open air or in enclosed vessels with controlled environments, allowing organic waste to decompose and break down for use as agricultural or garden compost.

anaerobic digestion – a biological process similar to that used for sewage sludge, in which biodegradable waste is enclosed without oxygen and broken down by bacterial action to make compost and generate bio-gas.

gasification and pyrolysis – similar techniques that apply extreme heat to carbon-based waste (paper and plastic, for example) and organic material to produce ash (or char), oils and a synthetic gas known as syngas which can be used to generate electricity and/or heat. They are examples of 'Advanced Thermal Treatment' (ATT).

ethanol production – fermentation process, similar to that used in beer and wine-making, which converts waste into liquid fuel and leaves a stable solid residue. The technique currently is used mainly with agricultural 'bio-crops' grown specifically for the purpose, but a plant designed to burn municipal waste is being built in the USA.

mechanical and biological treatment (MBT) – hybrid system in which residual waste is mechanically processed to remove any metals, glass, plastics and contaminants (batteries, for example) that still remain after kerbside sorting. The biodegradable residue is then composted. Where this is by anaerobic digestion, it will create methane to power the plant. The residue may then be either landfilled or used to make 'refuse-derived fuel' (RDF) for use in cement kilns and power stations.

bio-mechanical treatment (BMT) – similar hybrid system in which waste is shredded and dried before sorting.

Given the political impossibility of huge, mass-burn incinerators, it is the hybrids that offer the likeliest way forward. They have numerous advantages, not least of which is their long record of success in other countries – principally Germany and Austria, but also Switzerland, Italy and the Netherlands. They can work well in conjunction with kerbside recycling schemes, separating out recyclables that have been wrongly included with the general waste. This reduces the volume of residual waste and thus also the size of landfill needed to absorb it, and the local authority's disposal costs. It also makes landfill safer. Hazards such as batteries, fluorescent light tubes, paints and solvents are removed at the sorting stage. Being less biodegradable, the waste is much less likely to cause either liquid leakage or a build-up of methane; being more stable, it reduces the risk of dust, smells and wind-blown detritus. Unlike the vast, mass-burn incinerators, the plants are made up of small units that can increase or reduce in number as the type and quantity of refuse changes over time; and they can be built and operated on a small, local scale.

These are genuine benefits, but they do not come without

risk. Friends of the Earth fears that some local authorities will adopt MBT only to misuse it. They will see it as a 'means to meet recycling rates without the need for the separate collection of recyclables. But the dry recyclables separated out during the [MBT] process will be of poor quality compared to that collected by kerbside or bring-bank schemes.' There is also the problem of long-term contracts. If MBT plants require fixed tonnages of waste, then they could undermine recycling and other more desirable waste minimization schemes. This would apply particularly where plants produce RDF for cement kilns. The fuel's high calorific value depends on heavy inputs of plastic and paper, which – if the UK's adoption of the European waste hierarchy actually means anything – would be better recycled. Neither is there any guarantee that the improved performance of MBT over older technologies would be properly rewarded in legislative terms. Although the process greatly reduces the biodegradability of what is sent to landfill, it is still not totally inert and thus may not count towards EU Landfill Directive targets (though there is hope that this apparent anomaly will be resolved by a new directive in 2005).

All this greatly oversimplifies not only the technologies but also the economic, environmental and political quagmires through which they have to tread. If Donald Rumsfeld were to redirect his energies from war to waste disposal, he would find plenty of scope for his three degrees of uncertainty – known knowns (things we think we know), known unknowns (things we know we don't know) and unknown unknowns (things we don't yet know we don't know about). The battlefield analogies are hard to escape. There is a need for clear objectives, strategic planning and effective hardware. There will be gains and losses; occasionally unwilling sacrifices; collateral damage. There will be enemies. With luck, there will be a negotiated peace. More than any previous war, however, the battle with our own baser instincts is one in which appeasement has no place. It adds new meaning and sharp piquancy to the famous words of William Learned Marcy to the US Senate in 1832: 'The politicians of New York . . . see nothing wrong in the rule, that to the victor belong the spoils of the enemy.'

It has to be said that as members of the public we do not set our legislators the best of examples. Some years ago, while rummaging for a glove in an Icelandic desert, I accidentally spilled a sweet wrapper from my pocket. The wisp of paper blew away, a tiny diminishing flake in the dark expanse of lava, and I ignored it. Not so my local guide. He raced after it, following its every twist in the wind like a cheetah after a gazelle. When eventually he managed to trap it, he stowed it solemnly inside his anorak and walked back with a sad little shake of his head.

'No!' he said. 'We don't do that here.' By 'that', he meant behave like an Englishman, with a stubborn disregard for anything but his own convenience. It struck me at the time as a preposterous piece of dramatized pomposity. The wrapper itself was *prima facie* evidence of its unreasonableness. The only reason it was in my pocket was because I don't drop litter; I take it home. Quite apart from which, the impact on the Icelandic landscape would struggle to equal that of a fly-dirt, and who needs lectures on environmental good manners from a nation of whale-killers?

As acts of aggression go, however, over-meticulousness is

hardly in the capital class. In the Icelander's case it did at least express, in exaggerated form, some kind of national ethic. When it comes to defining the national ethic of the UK ... well, at least no one could accuse us of primness or vain excess of pride. No hill or valley of the sceptred isle is uninvaded by litter. Even remote mountaintops are scarred by the leavings of climbers whose sense of achievement, and pleasure in the open air, are unimpaired by dumped drinks cans, gas canisters, food packaging and faeces. Everywhere else, with a lethargy characteristic of abandoned hope, the throwaway society crudely lives up to its name. Good intentions, like last year's election posters, are blown to rags.

One multinational US manufacturer has a web page boldly entitled 'Save the Planet'. (Well, why not? Someone's got to do it.) 'We're doing our part,' it says, '. . . by conserving energy, recycling and reducing the amount of trash we make.' Every packing case it uses is made from at least 50 per cent recycled materials. Its vending machine trays are 100 per cent recycled; its printing plant has been so clever at recycling paper, fibre, wooden pallets 'and other stuff' that it has cut waste by 65 per cent. Its scientists have been busy finding ways to reduce the weight of packaging without compromising the freshness of the product. In its home town, Chicago, it rewards workers who use public transport instead of cars. You could say it has done for the environment everything it could possibly do except abandon its own product.

The name of the company – Wrigley – is known in every part of the world that has ever heard the voice of an American. Even in China, average personal consumption is between 15 and 20 sticks of gum a year. Cross the strait to Taiwan and it goes up to 90–100. In the UK (where Wrigley occupies a

219

forty-five-acre manufacturing site in Plymouth) it is a jaw-testing 120–130. In America it is 180–190. But chewing gum is much older than its American accent. The Ancient Greeks liked to chew a kind of gum made from tree resin, and some researchers even believe they have found evidence from the middle Stone Age, 9,000 years ago, of gum made from birch bark tar – a substance which, unlike the more widely used betel nut and tobacco leaf, seems to have had no value as a narcotic and must have relied for its popularity on the sheer joy of mastication. Chewing gum already existed in America when William Wrigley Jr, a former soap salesman then trading in baking powder, came up with a unique special offer. With every tin of baking powder a customer bought, he would give away, ABSOLUTELY FREE!, two packs of gum. Wrigley may or may not have been a marketing genius, but he was sharp enough to recognize an opportunity when he saw one. When it turned out that more people wanted the gum than had a use for the baking powder, he did not hesitate. Wrigley's Juicy Fruit and Spearmint both hit the market in 1893, and everyone knows what happened next.

By the early 1940s, gum was a $6.5m dollar market and as American as apple pie. Its issue to GIs (whose teeth it was reckoned to clean, whose thirsts it quenched and whose minds it was supposed to keep focused) ensured its worldwide spread as negotiable currency, multilingual ice-breaker and love token during the Second World War. Sales rapidly quintupled, and went on mounting. In the five years from 1998, there was still enough slack left in the market for Wrigley's sales in the US and Europe to increase by more than a third. And by now it was not just the gum-chewers who were spitting.

In 2003, the second annual Local Environment Quality

Survey of England (LEQSE), commissioned by Defra from ENCAMS, reported that trodden-in chewing gum was, literally, the most tenacious scourge of the urban environment. 'By a large margin,' ENCAMS said, it was 'the most widespread source of staining in England' – no small achievement when you reflect that the previous section of the same report was about dog-fouling. Wherever you look, the ground is pocked with gum – showing up as dark or light blotches depending on the colour and condition of the paving. ENCAMS found that 94 per cent of major shopping centres were disfigured by it; so were 66 per cent of local retail areas and 71 per cent of bus and railway stations. Overall, 66 per cent of ENCAMS's 10,000-plus survey sites were affected, putting gum way ahead of the next worst cause of environmental staining, oil from motor vehicles. Even on roads, oil could achieve only the narrowest of margins over gum, and everywhere gum was by far the more difficult to shift. No longer is it made in the old-fashioned way from *chicle* – the resin of sapodilla trees from the ancient forests of Mexico. No forest on earth could keep up with the demands of a world population whose jaws never stop. Like car tyres and glues, modern chewing gum is made from synthetic rubber. Unlike tyres and glues it is treated with softeners, sweeteners and flavourings. Exactly like tyres and glues it does not biodegrade. As the Parliamentary Office of Science and Technology (POST) put it in a two-page advice note published for the information of both Houses of Parliament: 'Synthetic rubbers are stretchy, retain their properties indefinitely under all weather conditions, are resistant to aggressive chemicals and have strong adhesive properties.' In practical terms, what this means is that when gum hardens it becomes all but immovable.

It is estimated that three and a half billion gobbets of the stuff are spat out every year in the UK – most of them sprayed with a randomness limited only by humankind's tendency to congregate. Wherever people gather – around pubs, clubs, schools, swimming pools, cinemas, stations, taxi ranks, shops – the gum-pats rapidly spread and overlap like a tacky vinyl overlay. In Oxford Street the number of individual gobbings visible at any one time has been calculated at 300,000, and the average gum density in the busiest parts of Westminster is twenty per square metre of pavement. Nationally, 75 per cent of the UK population are reckoned to be gum-chewers. Cleaning up after them – at a rate of between 45p and £1.50 per square metre – costs local authorities £150m a year (de-gumming Trafalgar Square alone cost £8,500 in June 2003). Even this is likely to be an under-estimate. The cleaning process itself – usually hot or cold high-pressure jet washes, but occasionally manual scraping, steam-cleaning or even cryogenics (freezing the gum) – often damages the paving stones or tarmac, and does nothing whatever to dissuade the spitters from coming back. Six months is all it usually takes for the blank canvas to gather a new layer of impasto, and many city centres have to be cleaned several times a year.

The law, too, is a sticky wicket. A particular problem is the lack of any clear legal definition of 'litter'. Is chewing gum included, or is it not? Some local authorities – Leicester City Council, for example – believe it is, and have told their street wardens to issue £50 on-the-spot fines. Others are not so sure, and public opinion is double-edged. No one likes mess, but hefty fines for dropping gum ('when child-killers go free!') are an open invitation to tabloid headline writers for whom political correctness is forever going mad. Defra nevertheless is determined to persuade others to follow Leicester's example, and says it is prepared if necessary to amend the legislation to make it clear beyond all doubt that gum is legally litter. Borrowing the mantle of environmental campaign groups, it also reminds citizens that the Environment Protection Act of 1990 gives them the right to take legal action where 'acceptable standards of cleanliness' are not enforced.

And it doesn't stop there. Mindful of the 'polluter pays' principle that underpins European waste policy, Defra has its eye on the manufacturers too. What are *they* going to do about it? If the likes of Mercedes and BMW will have to be mindful of what happens to their products when they expire, why not the *soi-disant* planet-saver Wrigley? One solution might be to change the chemical structure of the gum base, to make it biodegradable. If it were less sticky, spent gum could simply be swept up with other litter. If it were biodegradable it would melt away over time like apple cores or spilled chips. Defra is not optimistic, its frustration made only too clear in the POST paper of 2003:

> However, the gum base also determines commercially important features of chewing gum such as flavour retention, chewiness and shelf life. The challenge is to develop a non-sticky or biodegradable gum base that does not compromise the other features. Manufacturers are reluctant to release details of their research programmes for reasons of commercial confidentiality. However, it is generally assumed that little progress has been made. Wrigley's says that it has invested some £5m on research in this area in the last five years but no new products are yet ready for consumer testing;

other manufacturers say they see little incentive to invest in this area because there is no obvious financial return.

If the gum itself cannot be made less sticky, mused POST, then perhaps the pavements could be given some kind of non-stick surface to make the gobs more easily removable. One cannot know whether foreknowledge of such a development would have caused William Wrigley Jr to flush with pride or with embarrassment, but it seems a high price to pay for a baking powder promotion. And who, indeed, would pay it? Several local authorities already have told Defra they want to see clean-up costs underwritten by a levy on manufacturers; and the Irish government has been openly discussing the possibility of a chewing gum clean-up tax. No one, yet, seems ready to follow the example of Singapore, which imposed a complete ban on chewing gum in 1992. This held until 2002, when sugar-free gum was made available on prescription following a trade deal with the US. Defra did toy with the idea of voluntary sales bans in city centres, or other areas where the problem was especially acute, but POST saw this as a non-starter: '[Manufacturers] argue that this is unlikely to have much effect because few people buy, chew and dispose of their gum all in the same area. Further, [local authorities] have expressed concerns that such voluntary schemes would be unworkable.'

Some local authorities have gone their own way. Bournemouth, for example, was one of several to try GumTargets – the idea of an inventor who noticed people sticking gum on posters as they rode up and down the escalators on the London Underground. Boards hung on Bournemouth lamp-posts bore inviting pictures of famous people whose likenesses, it was thought, would invite defacement by discarded gum-blobs (Jeffrey Archer, Sven-Goran Eriksson, Tony Blair, Iain Duncan Smith, Jordan, Wayne Rooney and Kermit the Frog were among those tried). Bournemouth reported an average of 1,600 pieces of gum on the targets a week, with 'a significant decline in chewing gum litter on pavements', but eventually abandoned the idea when, after the novelty wore off, people resumed spitting on the ground and it became clear that, with or without boards, the paved areas would still have to be cleaned.

As a mirror held up to the nation's sense of self-worth, LEQSE reveals a curious inversion of behavioural norms. Instead of erasing blemishes, we multiply them. In just a year, the volume of fast-food rubbish discarded in England went up by 12 per cent. Thanks to 'Drive-Thru', the area affected by each outlet now extends over many miles, and takeaway-eaters show little more inclination than mountaineers to take their leavings home with them. Burgers, kebabs and pizzas add their special gloss to the gummy underlay, and are sauced in their turn by vomit and urine. Because fast food is mostly dropped in the evening, long after street-cleaning teams have gone home, it remains an essential part of the image that towns and cities impress upon their visitors. The rubbish and graffiti around London's Underground, bus and railway stations are the worst in the country and, in ENCAMS's words, 'leave visitors with the impression that London is the dirty man of Europe'.

'The simple fact is that we're not being canny enough,' said Justin Jupp, Keep Britain Tidy's regional director for London. 'In cities such as Brussels, Frankfurt and Vienna, they concentrate their efforts on tourist hotspots and littered grot-spots, ensuring that visitors leave with the notion that the city is clean. They are also more strategic in their planning and flexible in their approach, and have adapted to the twenty-four-hour society – something that some London Boroughs still haven't come to terms with.'

This in its way is as dismal a testament as any that could be imagined – as if street-cleaning were a matter of touristic window-dressing designed to create a 'notion' of cleanliness, and as if out of sight were out of mind. The twenty-four-hour issue, however, is a real one. 'In England,' complains ENCAMS, 'many councils still send their sweepers out during the day, when they can't do their job properly because people, signs, café tables and parked cars are in the way.' Nationally, it found, 70 per cent of streets and gullies were not cleaned properly because of obstructions. 'Teams often went out in a mechanical sweeper, which can't reach trash stuffed behind GPO boxes, or had a brush when the pavements were blighted with grease, oil, sick and gum.' On only five of the thirty-eight occasions on which researchers watched cleaners in action was the result marked as satisfactory.

Overall, the LEQSE report measures standards across eleven different environments, from industrial estates to private housing and rural roads. As well as litter and staining, it looks at weed growth, fly-tipping, fly-posting, litter bins, lamp-posts, street furniture, bus stops, public lavatories, graffiti, traffic and pedestrian flows – anything at all that can be used to measure the impact of a place on the people who live or work there, or who visit or pass through it. The most encouraging thing it has to say is that the margins of failure are often narrow. Many places only just fall short of basic national standards, and would need only an injection of will to bridge the gap. Fall short they do, however, and will is hardly evident in what almost everywhere is a worsening trend.

High-density housing comes off worst. Seventy-one per cent of terraced housing, flats and maisonettes, and 68 per cent of urban social housing, are in areas classed by ENCAMS as 'unsatisfactory'. The effect on the national psyche is cumulative, chronic and degrading. With physical decay comes spiritual blight, as pride and optimism follow social responsibility into the shredder. Out of despair and cynicism is born a new kind of pragmatism whose best expression is a shrug. So who cares? Get real! Apathy and powerlessness leach from the population into the administrative fabric of local government, headquarters of a new political realism that often fails to accept even that statutory responsibilities are achievable.

'It is clear again this year,' said ENCAMS, 'that over two thirds of the population that live in higher density or social housing are living in an environment which is not meeting the statutory requirement. Put bluntly, local authorities and others are not meeting their obligations.' The environmental quality minister, Alun Michael, joined the chorus:

It is my strong belief that there is a continuum between the way people treat their public spaces and the way they behave towards one another, linking environmental issues like litter, which pull down the feel of any public space, with anti-social behaviour, vandalism and violence. We cannot create sustainable communities without ensuring that the standard of our public spaces is up to scratch.

Encouraged, perhaps, by the fact that the dominant colour on the ENCAMS chart was yellow (for 'unsatisfactory') rather than red (for 'poor'), he had a stab at optimism. 'The report shows simple management changes by local authorities could improve the quality of local environments – at no extra cost to the taxpayer.' Thus unwittingly was the cynic's view made official. It's not money that's the problem, it's attitude. The minister's conclusion – 'We don't have a right to clean streets unless we take responsibility for our actions' – is a smooth reiteration of the old Wrigley classic, dating from 1933: 'Use this wrapper to dispose of gum'. The circle, alas, is all too easily squared. The place is in such a mess already – litter not cleaned up, street furniture and lamp-posts not maintained – what difference does it make if I spit where I like?

Other environments fared little better. Sixty-six per cent of local shopping and business areas were rated unsatisfactory by ENCAMS; so were 63 per cent of bus and railway stations, 63 per cent of industrial estates, alleyways and out-of-town shopping areas, 62 per cent of main roads, 55 per cent of urban high streets, 50 per cent of parks, picnic sites, canals, lakes, riversides, seafronts and promenades. Only suburban housing (52 per cent judged satisfactory or better) and rural roads (54 per cent) came out nationally on the right side of the balance – which of course means that 48 per cent and 46 per cent didn't. The report's authors did their best to emphasize the positive – improvements in dog-fouling, graffiti and the removal of fly-posting – but their cheery encouragements ('Thumbs up for the managers who look after our town and city centres'; 'Keep up the good work if you want to see the English seaside boom again'; 'A continued improvement. Keep it up!') seemed less like celebrations of victory than the ever-hopeful, ever-loyal touchline chorus from the fans of Rubbish United.

CHAPTER SIX

Poison

A FAT GREEN GOOSE-DROPPING LIES IN THE MIDDLE OF THE path. Not far away a drowned sheep, martyr to its own stupidity, rots in the sun where the tide delivered it. To the south is a freshwater mere, upon which geese and ducks ride motionless like fleets at anchor. For a moment the tableau freezes – like a diorama, almost unreal in the sharpness of its detail. But peace in the wilderness is never perfect. Every sinew is tuned for survival; every appetite insatiable; every take-off and landing a cue for trouble. Territories are infringed, mates eyed up, meals stolen. Antagonists collide breast to breast in squalls of feather and spray. Life here is stripped to its fundamentals. A sexual assault by a drake on a duck ends with both of them taking to the sky and arrowing landward, dead straight, as if she hopes to outrun him (or perhaps, more cunningly, to lead him somewhere). Other confrontations are so stupid they are almost human. Two herons decide the mere is not big enough for both of them and totter away like elderly drunks flapping their coats.

Shadows harden as early summer turns up the heat. Waders

wade; damselflies dart; reed buntings flit and cling. To the north, a pair of avocets with their fragile-looking, upside-down beaks spear their reflections in the creek. Swans sail in the half-tide on private oceans of contempt. Beyond the creek is marsh, hunted by owls and harriers, then mud and miles of tide-rippled sand that fudges the line between land and sea. On the furthest sandbank, appearing from a distance like fat black maggots, are the seals of Blakeney Point. To the west, beyond a grey smudge of pine, lies the shallow dish of Holkham Bay, rimmed by dunes and trees; then the tight little harbour at Burnham Overy Staithe where weekending Londoners, all soft city faces and gnarled legs, wrestle with the muscular intricacies of sail. In such places European Waste Directives, recycling targets and Landfill Tax Credits can seem as distant as the Iraqi war. On holiday weekends and fine summer evenings there are occasional intimations of threat: 4x4s on the coast path, a couple of howling jet-skis off the point, mis-directed migrants from the urban junklands, unable to feel pleasure unless they are burning fossil fuel. But they are exotics who will not endure. Here, on the hip of England, is the natural heartland of zero waste. Here lives end, fuse and become life again. Goose-droppings, avocets, reeds, grasses, marsh plants, wild flowers, razor clams, seals, sheep – all enrich the compost in which the future will implant itself and grow. It is the deepest of ironies that we find the greatest uplift in places least touched by our own species. But here also is the most powerful antidote to apathy. England may have poison in its veins, but there is vigour in the body yet, and beauty. There may be little glamour in what has to be done, and you may not care for the beardy-weirdy nature of certain of your fellow travellers, but let no one say they can't see the point.

Such thoughts are not without risk. They can lure you like wreckers' lamps on to the reefs and shoals of over-simplicity, washing you up in *Thought for the Day* country. Too easily you find yourself grasping at words like 'natural' or 'synthetic' as if they had clear and precise meanings, and somehow stand for good and evil. For what defines 'nature' if not the immutable laws of physics and chemistry? If yoghurt must obey these, then so too must nuclear power stations, mobile phones and moon rockets. If they didn't, they would be like cold fusion, alchemy or time travel – the soft furnishings of a dream world. Is margarine 'natural'? Is tofu? Is a fleecy-lined jacket made from recycled plastic bottles? Is aspirin? Are clothes? Cooked food? Bicycles?

Of course it's the wrong question. Natural versus synthetic is as much cholera versus antibiotics as it is fresh air versus dioxins. In varying degrees of complexity, these are matters of philosophy and of its homespun passenger, common sense. One aspirin a day may prolong your life; thirty will end it. Low doses of sunshine make vitamin D; high doses make melanoma. Few things are absolute. The 'natural' toxins in organic food are as likely to cause cancer in laboratory rats as are the 'synthetic' ones in conventionally farmed fruit and vegetables. Words like 'poison', 'toxin' and 'carcinogen', too, need careful qualification. Language itself is a big part of the problem. It is the opacity of its vocabulary as much as its physical complexities that places science beyond our understanding. Science makes everything sound scary, and so we are scared. Even that neutral word 'chemical' has been tainted by the company it keeps. Consider, for example, acetaldehyde methylformylhydrazone, allyl isothiocyanate, crotonaldehyde, 1-hydroxyanthraquinone, monocrotaline, quercetin and symphytine. Who knowingly would swallow any of those? All are pesticides. All are carcinogenic to rats. Yet we all will have ingested them, or various of their relatives, in apples, bananas, broccoli, cabbage, carrots, celery, coffee, garlic, grapefruit, grapes, honey, lemons, lettuce, mushrooms, onions, oranges, parsnips, peaches, pears, peas, potatoes, raspberries, tea, tomatoes . . . The list would fill the page. Coffee alone contains more than 1,000 different chemicals, of which (as I write) only some twenty-seven have been assessed for carcinogenicity. Nineteen of these have tested positive. No one knows about the rest.

And yet there is no escaping them. Toxins like these are all produced within the plants themselves as part of their evolved defence against fungi, insects, birds and animals. Conventionally grown, organically grown, picked from the garden, it makes no difference. They are *natural pesticides*, having no other purpose than to be poisonous. Ninety-nine point nine per cent of the chemicals we swallow are natural, and are found in fruit, vegetables and bread. They have existed throughout the entire evolutionary progress of vertebrate life, but still cause cancer in mice, rats and – well, who knows what else? Literally thousands of such substances are in free circulation, of which only the tiniest fraction – a total at the last count of just seventy-one – has ever been tested.

Day by day the natural pesticides in our diet outweigh the synthetic by 20,000 to one – the approximate scale of difference between an elephant and a mouse. In terms of toxicity, so far as anyone can tell, there is little to choose. Two thirds of synthetic chemicals are carcinogenic to rodents; so are two thirds of the natural ones. Scientists at the National Institutes of Environmental Health at the University of California at

Berkeley calculated that the known natural rodent carcinogens in a single cup of coffee are about equal in weight to the synthetic pesticide residues in an average American's entire annual intake of fruit and vegetables – a figure made all the more striking by the fact that only 3 per cent of the natural chemicals in coffee have been tested. Every day, each of us consumes some 2,000mg of carcinogenic or mutagenic material – an approximate quarter teaspoonful – created by cooking. Over the same period, according to the US Food and Drugs Administration, the average consumption of synthetic chemicals is just 0.09mg – about the weight of a single grain of refined salt. And yet it is the synthetics that ring the alarms.

What should we deduce from this? That we have been encouraged by environmental propagandists to become a generation of chemophobes with an irrational dread of science? That the statistics serve only to lay a smokescreen around the chemical industry, disguising the fact that our bodies, through long exposure, have developed a resistance to natural toxins that is no protection against the small but lethal minority of laboratory products? That there is everything to fear? Or nothing?

Pesticides are the most obvious example, but there are plenty more. Almost all natural processes produce hazardous wastes (think of your own body), and so do most industrial processes. We have to mind our language, to recognize that 'chemical' is no more pejorative a term than animal, vegetable or mineral, and to accept that the existence of 'chemical waste' is not of itself an argument in favour of abolishing the process that creates it. Understanding this is a prerequisite for any sensible argument about the creation, control and disposal of hazardous wastes. Sensible arguments, however, are scraps in the wind if they are not heard and acted upon. To the immense frustration of other 'stakeholders' – civil engineers, local authorities, environment groups, waste disposal companies – we stumble yet again across the inert form of the UK government. It knew what was coming; it had ample time to make ready; and yet it held only a paper parasol against the poison rain.

The key date was 16 July 2004. On this day the EU Landfill Directive brought radical change to the way hazardous waste was to be defined, treated and disposed of. Broadly speaking, 'hazardous' or 'special' waste is material liable to cause death, injury or illness to people who touch it, or that can damage the environment. Until the summer of 2004, England and Wales annually produced some 5m tonnes of such waste, of which a little under half went to landfill. At midnight on 15 July, however, everything changed. It was no longer lawful for hazardous and non-hazardous wastes to be mixed in the same landfill. As the clock struck twelve, the number of sites allowed to receive 'special' material fell from about 240 to fewer than ten. In some areas – the West Midlands, for example – it brought the total down to zero. At the same time, the EU added another 250 names to the list of substances classified as hazardous and requiring special treatment. It was double-whammy time again. As the volume of special waste went up, so the number of treatment sites was reduced by more than 90 per cent.

In an ordered world, the government would have done everything necessary to prepare for the changes – unambiguous policies, clear guidelines, new infrastructure, whatever it took. But Westminster is not an ordered world. It is a palace of torments in which humble virtues fall between the furious assaults of political rank-pullers and the clogged arteries of an administrative process stiff with committees. Of course the government had due warning. The EU Landfill Directive became law in July1999, giving us exactly five years to prepare for July 2004. There is a school of thought that believes the UK, in lacking the experience of its European neighbours, was simply naive. 'It's quite possible,' says Biffa's Peter Jones, 'that the civil servants didn't know what they were agreeing to. We are now paying the price for that.' If ignorance (the defence of fools) is better than lethargy (no defence at all), then perhaps this view is the more charitable one. But harder heads in local government and the waste industry knew a storm cloud when they saw one, and were quick to raise the alarm. In 2002, for example, the National Household Hazardous Waste Forum told the Environment, Food and Rural Affairs select committee: 'The government shows a lack of preparedness in recognising and responding to the requirements of impending EU legislation. Those likely to be affected . . . need clarity and decisiveness when new legislation is posed, not the delays and obfuscation that seem to be the norm.'

The Chemical Industries Association (CIA) agreed:

Our underlying message is that uncertainties . . . make it difficult for waste producers to develop coherent medium to long-term strategies. Whilst this situation persists, it is difficult to envisage that, in the short-term at least, waste producers will truly be able to take responsibility for planning the final destination of their waste. In the meantime, our members' level of concern is increasing . . .

The chairman of the environment committee's hazardous waste sub-committee, Michael Jack MP, heard them loud and clear. Questioning the CIA's Director General, Judith Hackitt, he articulated the fears of the entire industry. 'One of the things that concerns me,' he said, 'looking at this picture you are painting of a lack of forward planning, is, are we heading for a chaotic situation where, because of this lack of definition and the timetable looming, we will end up with piles of chemicals here, there and everywhere, and people saying, "What do we do with them? How do we dispose of them? What is the guidance?" Is that too exaggerated a position or not?'

Ms Hackitt could offer no comfort. In the absence of a proper solution, she said, companies producing hazardous waste 'will have to make decisions about what to do . . . and one of the possible solutions is to continue to store it'.

The waste management contractor Cleanaway made no attempt to disguise its irritation:

> We do not know what criteria waste will have to meet to be allowed into landfill, we do not know what level of treatment will be required and we do not even know the exact dates when parts of the Directive will come into effect. Government cannot leave the future of hazardous waste management to be determined by market forces and then bemoan the fact that businesses arc cautious of investing before knowing whether there is a commercial case for it.

The select committee's report was published on 17 July 2002. Its twenty-two conclusions and recommendations, listed alphabetically from (a) to (v), were surprising only in that anyone should have thought them necessary. Recommendation (b) concluded with the plea: 'We can only reiterate our previous recommendation ... that ... any new item of European legislation should not be agreed until all the practical implications of implementation are well understood.' In other words, comprehension should precede signature, not vice versa.

You have to remind yourself that this homily is directed not at a class of eight-year-olds but at government policy-makers. Recommendation (e) explains the importance of knowing what you're talking about. 'The Environment Agency and DEFRA,' it says, 'must work with the waste management industry to provide timely high-quality data on the amount of hazardous waste produced each year and to develop management methods to assist in planning for future capacity.' Did anyone ever think otherwise? Recommendation (i) is such an historic favourite, it could be stencilled on the committee's T-shirt: 'The Government should clarify its position on the specific role of incineration in the disposal of hazardous waste.' Just in case anyone thought policy could be left to the man who services the photocopier, (j) offers a terse reminder of where the buck stops: 'We recommend that the Government takes the utmost care to ensure that such consultations occur as early as possible, are of the right kind and are at the right level.'

By the time it reaches (r), the committee has begun to sound like a worn-out schoolmaster wearily demanding the truant's homework: 'The Government must make clear what specific targets, if any, it has set ... and what positive steps it has taken to achieve those targets.' In (v), all hope gone, it collapses into desperate reiteration of what everyone knew before the whole thing started: 'What is clear is that the Government and industry must form a partnership for the management of hazardous waste to ensure that, in 2004 and beyond, we have an adequate and environmentally appropriate hazardous waste management infrastructure.'

It cannot be said that the government ignored the advice entirely. At the height of the storm, amid all the terrifying calls for action, it was thrown a lifeline. Recommendation (u): 'The Government should encourage the development of a national hazardous waste forum to address the issues . . .' It grabbed with both hands, offered a silent prayer to the benevolent gods of *mañana*, and let its worries float away. A talking-shop! Yup! Just the ticket!

By December 2002, three years and four months into the Landfill Directive and just nineteen months short of the 2004 deadline, the members of the Hazardous Waste Forum were appointed. Their brief: 'to consider the demands on industry of existing and forthcoming legislation, to consider targets for hazardous waste reduction and recovery and to provide a means for bringing all relevant sectors together to work towards the goals of reducing hazardous waste and managing it safely'. In effect, they were starting again from scratch. The membership came from central and local government, official regulators, producers and handlers of waste. They met for the first time in February 2003 and produced their 'action plan' in December the same year. By now the country was another whole year down the track, with just seven months left before the engine hit the buffers.

It would be unfair to suggest that the forum did a bad job. Its members were well qualified, and did as much as they could. They applied themselves conscientiously and came up with a detailed schedule of proposals which – though coming nowhere near the kind of zero waste policy that would satisfy the idealists (like the select committee, they foresaw a major role for incineration) – at least met practical questions with practical answers. But they did not try to reinvent the wheel, and there were lingering echoes of every other traveller who had passed along the way. Their first recommendation: 'Government should provide clarity as soon as possible on the extent, timing and application of forthcoming changes to legislation governing the management of hazardous waste.' Well ... yes. Sound familiar? Other recommendations, too, were worn smooth by repetition:

Clearer information setting out likely capacity requirements should be prepared for the hazardous waste recycling, recovery, treatment and disposal capacity, which needs to be developed in the short and medium term, to meet the requirements of the new regulatory control regime ...

The Forum considers the use of targets for hazardous waste reduction as an important means of driving and monitoring progress. However, the information required to set realistic and achievable targets is not currently available ...

The Forum sees the need to provide, and maintain in the long term, a diverse range of options for hazardous waste recycling ...

Separate collection of household hazardous waste should be encouraged ...

And so on. No one could argue with any of this, and no one did. Speaking on the very day the plan was published, Dirk Hazell, chief executive of the Environmental Services Association, one of the forum's own members, readily acknowledged its value. It had, he said, 'identified many of the key issues that need to be addressed'. And yet it was all too cobwebbed and mired in delay.

> For years ESA has warned that the UK risks a hazardous waste crisis from next July: even now, the UK's waste managers still do not know how the government intends to manage hazardous waste from July so it is not possible to invest in treatment plants. Indeed, the authorities' failure to enforce regulation has forced our members to close modern plants which manage hazardous wastes in the safest way…The authorities will have to use the remaining time much more productively than they have so far shown themselves capable of doing if they are to avoid sleepwalking the whole country into a hazardous waste crisis from next July [2004].

Being anything but fools, members of the forum knew very well that, even if the government were now to awake and pursue their every suggestion, the policy vacuum could not be filled. They were five years too late. Such was the certainty of failure that they enshrined it in their own recommendation: 'Defra and the Environment Agency should consider what short-term actions may be required in the event of a hiatus between the implementation of Directives … and the pro-vision of appropriate management options. This might include a general shortfall in treatment capacity, lack of an appropriate management route for a specific waste stream etc.'

The only misplaced word here was 'might'. Without new infrastructure, which there was no time to plan and build, the UK had no chance of being able to cope. 'If the treatment is not in place in time,' said the forum, '. . . there is the potential for large amounts of hazardous waste being produced with no treatment or disposal outlet.' There hardly seemed a need to spell out the consequence but it did so just the same. 'This has serious implications for the environment, for the sustainability of UK industry and for the country as a whole.' And so it came to pass. The treatment was not in place; hazardous waste joined the ranks of the homeless, and 'the environment' braced itself for the next dose of poison.

Paragraphs from official documents make dull reading, but they are important markers in the unfolding story. If Lord Butler had investigated the quality and use of intelligence in the UK's war on waste, then his conclusion might have been similar to the one he reached on Iraq – that there had been a colossal dislocation between the evidence and the action. The difference in this case was that the pendulum had swung the other way. Instead of sexing up we got damping down: bromide instead of Viagra. The Hazardous Waste Forum's action plan was a fig leaf on a eunuch, with a potency of zero.

As late as June 2004, a month before the changes were due to bite, the Institution of Civil Engineers' spokesman in the West Midlands, Peter Braithwaite, a director of the civil engineering giant Ove Arup, was still pleading for a miracle: 'We are not scare-mongering. This is a problem that will not go away . . . It will need a co-ordinated drive . . . We urge all parties to work together.'

When 16 July came, the situation looked something like this. ('Something like', because there were almost daily swings in the estimated number of registered landfills.) Mixing hazardous and non-hazardous waste in the same landfill was now illegal. Landfill sites had to be registered in one of three categories – inert, hazardous, or non-hazardous – and could accept waste only of the permitted type. *Inert* waste is exactly what it suggests, stable material that is neither combustible nor biodegradable. It includes, for example, concrete, bricks, tiles, cement, glass, and uncontaminated soil and stones (if it's contaminated, it's hazardous). *Hazardous* waste is the tricky stuff, exhaustively listed in the EU's newly extended Hazardous Waste List. It includes everything from contaminated soil to incinerator ash, asbestos, acids, slags, tars, oils, demolition and construction waste, industrial wastes, paints, fluorescent light tubes and television sets. If it were to be landfilled after 16 July, such stuff would have to be treated to reduce its bulk and make it safer to handle or recover. 'Treatment' in this context meant anything from physical sorting to chemical or biological processing, or burning. *Non-hazardous* waste is all the rest, including municipal garbage.

Not the least of the questions being begged was the influence of the market. Waste disposal plants, whether they are landfills, incinerators or any other, are privately operated and commercially driven. The owners are in it to make money, and are no more likely than any other businessmen to make risky, leap-in-the-dark investments. Neither do they go looking for trouble. How many of the 240 landfills receiving hazardous waste in England and Wales before 16 July would register to continue afterwards? Answer: nine, with four or five more considering their position. In huge areas of the country – the South-East, West Midlands, the whole of Wales – there would be none at all. What, then, would happen to the waste? Answer: that's a very good question.

One immediate effect was likely to be a huge increase in hazardous road miles. 'Waste will be travelling the country looking for a home,' said Peter Gerstrom, chairman of the Institution of Civil Engineers' Waste Management Board. Given that the overall total of hazardous waste produced in England and Wales is around 5m tonnes a year, this amounted to wholesale risk trafficking. The government-endorsed 'proximity principle' insists that waste should be disposed of as near as possible to its point of origin. Yet, here again, principle and practice faced each other across an abyss. You could argue that this was a benevolent evil; that the industry had to be forced away from landfill and made to face the consequences; that this was the only way to make it find better alternatives. By this logic the government was acquitted of inertia. Far from standing idly by, it was applying exactly the kind of regulatory pressure its critics said they wanted. At the cost of a little short-term nuisance, it would ensure a cleaner, safer future.

If only. The big stick is all very well, but it needs firm support from its operational partner the carrot. Whereas the industry could, more or less, be clear about what the government *didn't* want, there was no certainty at all about what was wanted instead. 'The industry can only work within the rules,' said Peter Gerstrom, 'and it needs those rules in advance.' This had been at the root of the industry's frustration for more than five years. In one sense, waste is as predictable as water – it will always find the easiest route. There is no incentive for companies to develop expensive, high-tech options if they are going to be undercut by simple, old-fashioned landfill. No one will make that kind of investment in a climate of uncertainty. There have to be tax advantages. The quantity and quality of waste has to be guaranteed. There has to be a policy framework of sufficient clarity and robustness to attract heavyweight, long-term investors. Playing fields have not so much to be levelled as tilted the other way. Until that happens the industry will sit on its hands.

On the day itself, the headlines were all about New Labour's by-election disasters, the fallout from the Butler Report and a pair of Union Jack trousers worn by a golfer at the British Open. Of impending disaster in the hazardous waste trade there was hardly a word, though on the previous day there had been some small coverage of Alun Michael's proposal to increase the penalties for fly-tipping. In the unreported hinterlands of the legitimate waste trade 16 July dawned wet, fearful and uncertain. In the cowboy trade it dawned full of promise.

There were two reasons why landfill companies did not rush to specialize in hazardous waste. The first was commercial. Five million tonnes sounds, and is, a formidable amount, but hazardous waste still accounts for only 1 per cent of the national total. Simple arithmetic dictated that the majority would opt for volume rather than specialization. The second reason lay in the same old graveyard of government policy-making. Separating waste into three discrete streams – inert, hazardous and non-hazardous – depends on precise definitions of what should go where, on rigorous checking and testing, and on the existence of treatment facilities for the hazardous stuff. In July 2004, none of this was in place.

The crucial what-goes-where issue was supposed to have

been set out in detailed, legally enforceable Waste Acceptance
Criteria to be policed by the Environment Agency. There was
never any chance that they would be ready in time – a fact that
was already evident when the Environment, Food and Rural
Affairs select committee reported in July 2002. Hence its all-
too-typical conclusions:

(a) This delay [in determining the Waste Acceptance Criteria]
should not have occurred.

(b) We are concerned that landfill operators are required to
make crucial decisions about the future designations of their
sites without the Waste Acceptance Criteria having been
agreed.

The industry's frustration boiled over in March 2004, after
it became clear that it would have to wait another whole year,
until July 2005, for the criteria to be put in place. The Environmental Services Association, whose member companies were
apparently expected to pull the government's chestnuts out of
the fire, came out blazing. 'The Landfill Directive was a done
deal by 1999 . . .' said its chief executive, Dirk Hazell.
'However, only today has the government made an incomplete
announcement about how it proposes to deal with hazardous
waste after . . . July 2004.' The government's failure to adopt
EU standards until July 2005, he said, would mean a period of
heightened risk in which safe, carefully regulated waste
management facilities would be abandoned for the twilight
world of illegal dumping. Upon the ministers and officials
responsible for this fiasco, he emptied a dustbin of scorn: 'It is
almost beyond belief that the authorities have still not bothered
to disclose their precise requirements for treatment of

hazardous waste . . . Given its track record since 2001 on such a basic environmental imperative as safe management of hazardous waste, what is the purpose of Defra?'

This is another good question without an easy answer. While preparing this chapter, I was offered a detailed briefing from a senior Defra official to make clear its policy and explain its thinking. The offer was accepted, but the meeting never took place and subsequent requests went unanswered. This made it difficult to form any clear idea of how this apparent absence of policy meshes with the department's mission statement: 'Defra works for the essentials of life – water, food, air, land, people, animals and plants.' By any normal reading of the situation, all these 'essentials' were being exposed to increasing risk. On the morning of 16 July, there were few outside the department itself who could see any likelier outcome than accelerated criminal activity, pollution and chaos.

In the short term at least, landfill operators would have to adopt what the minister of state for the environment, Elliot Morley, described to the House of Commons as 'a site-specific approach based on loading rates of new wastes, the types of new waste and the types of waste already in the landfill'. What this meant in practice was that the operators would have to take on trust their customers' accounts of what they were dumping, and it was the customers who faced prosecution if anything went wrong (always assuming, of course, that they could be identified and traced). Not until the Waste Acceptance Criteria came into force would legal responsibility pass to the site operators, who then would face the prospect of unlimited fines and a possible five years' jail if anything illegal slipped under their guard. This may be another reason why operators were in no great hurry to register their sites.

Lower down the feeding chain, where the hyenas lurk and there are no reputations to ruin, a run-in with the law is no more of a deterrent than any other minor occupational hazard. Crime likes nothing better than an under-supplied market, and will spread into the vacuum like weeds on to a bombsite. Many of the wheezes were already well known – hazardous waste hidden under top dressings of household garbage; sawdust soaked in poisonous effluent; contaminated soil passed off as inert; and the perennial favourite of the waste-disposal underclass, wait till you're out of sight, then tip it in a lay-by. 'There are plenty of unscrupulous operators in the waste industry,' says Peter Gerstrom. 'Guys will take a consignment and it will disappear.' Which means? 'It will turn up fly-tipped in some corner of the countryside.' With commercial pressures doing what they always do in times of scarcity – i.e., push up prices – the discrepancy between legitimate and cowboy disposal fees could only widen.

Fly-tipping and pollution were not the only risks. When one corner of the policy cloak comes unstitched, the entire garment may begin to unravel. In 2003 (the last year for which figures are available), the amount of previously built-upon, 'brownfield' land potentially available for redevelopment stood at around 66,000 hectares – 163,086 acres, or 254.82 square miles. Forty-five per cent of this might be allocated to housing – enough for 950,000 new homes, of which the greater share would be in the hard-pressed areas of London, the South-East and the East. Since 1998 the government has decreed that at least 60 per cent of new development must be on brownfield land, a target it has met every year since 2000. The economic importance of such land is obvious, and, despite the complaints of housebuilders for whom the only acceptable field is a green one, it is an equally important plank of environmental protection policy. On an island as crowded as ours, the last thing we can afford to waste is land.

The problem with brownfield sites is that they tend to be contaminated – with oil, metals, industrial waste, whatever their previous owners spilled or dumped on them. And this means, under the terms of the EU Landfill Directive, that soil and rubble cleared from the site is legally hazardous and must be treated as such. Peter Gerstrom's greatest worry was that higher prices and longer distances would lock the brakes of brownfield redevelopment and skid the builders back off into the green, or at least slow the process down. It will take time for the effects to work their way through – there was a kind of Klondike in reverse when developers flocked to the dumps before 16 July to beat the ban. 'They were quite cute,' said Gerstrom a couple of months later, 'and sent thousands of tonnes before the deadline.' With the summer surge still casting its shadow, it was not possible to quantify the full extent of the damage, but the signs were not good. Landfill prices for hazardous waste had not so much accelerated as warped into a different galaxy. In the South-East, where space was already running out and the few remaining operators needed to dam the gathering flood, the rate shot up from £10–£15 to £100–£200 a tonne. Everything went ominously quiet. 'The stuff just isn't appearing,' said Gerstrom. Some of it did take to the roads, but the volumes were small – nowhere near proportionate to the annual total of 5m tonnes – and the distances large. So what was happening to the rest? Much of it was simply being stockpiled by people waiting to see what would happen next. Gerstrom had heard a rumour that 3,000 tonnes was sitting on one development site alone. Gross that up nationwide and you have a geological feature big enough to worry air traffic control.

The remainder presumably was taking whatever route it could find, disguising its identity and following the old beaten pathway of nods, winks and crumpled money. Until this is blocked off, the incentive for legitimate site operators is reduced still further. 'There are issues of enforcement,' said Biffa's Peter Jones, 'because significant volumes of previous hazwaste have gone missing and the Environment Agency is unable to follow them up.' By late September 2004, none of Biffa's own landfills had been licensed to accept hazardous waste under the new regulations. As hazardous waste had accounted for approximately a third of what they received before 16 July, it seemed reasonable to expect that volumes of traffic would have dropped proportionally thereafter. But they hadn't. They had gone on precisely as before. So what was going on? Was it conceivable that the volume of inert waste had, by coincidence, on the same date, increased by an amount precisely equal to the hazardous waste that was now going elsewhere? No: pull the other one. Suspicion is not knowledge, and circumstantial evidence is not proof, but you'd need a cologne-soaked hankie over your nose not to catch a whiff of something rotten. Jones's conclusions were exactly the same as Gerstrom's. 'Apart from fly-tipping and stockpiling,' he said, 'the options are that small traders are packing it in black bin-bags, taking it to civic amenity sites in the backs of their cars and throwing it in with the household waste, and that hazardous building waste is being blended with harmless stuff and being sent out for recycling or to landfill as inert.' Who knows? The only certainty is that millions of tonnes of hazardous waste have vanished from the radar. 'The problem,' said Peter Jones, 'is that there is no central electronic database that tracks the movement of hazardous waste. No one knows what's happening. There is no absolute truth about what is leaving one place and arriving at another.'

In time, by one means or another, the situation will ease – though it may happen in ways that are not to everyone's liking. Nobody in the business perceives a solution that does not both perpetuate the use of landfill and increase the use of incineration. Talk of 'zero waste' will remain just that – talk. The Environment Agency, which licenses and regulates landfills, is encouraging owners to provide separate 'cells' for hazardous material that can be sealed off from the rest of the site. It is no quick fix. Each new cell, even when it is on an existing site, needs planning permission from the local authority as well as a permit from the Environment Agency – and nothing arouses fiercer opposition in a local electorate than the threat of hazardous waste on their doorsteps. The usual result is 'Toxic timebomb' headlines in the local newspaper and a mobilized populace egged on by environmental campaigners. Who can possibly blame them? The assurances of safety may be endorsed by every species of expert from civil engineer to seismologist, but decades of governmental ineptitude, false assurances and weasel words have worn away public confidence like acid on a doughnut. Nothing is trusted. Residents' action groups will fight up to, through, and even beyond public inquiries. The outcome may vary in terms of eventual winners and losers, but in one respect the result is always the same – delay, escalating costs and a disposal problem that grows by the hour.

One example often quoted is that of Britain's only working salt mine, near Winsford in Cheshire, whence comes the entire supply of rock salt for England's icy roads. In 1999, after two years of planning and consultation, a waste company called Minosus applied to Cheshire County Council for planning permission, and to the Environment Agency for a permit, to turn a worked-out section of the mine into an underground repository for hazardous waste, for which the cool, dry conditions were said to be ideal (local and national government records were already stored there). To government and the waste disposal industry it looked like a gift from the gods. The mine is over 550ft deep, and has been worked since the mid nineteenth century by the 'room and pillar' method that cuts away 70 per cent of the available salt, leaving 30 per cent in the form of pillars supporting the roof. The result is a vast honeycomb of high, empty rooms enclosing a total space of 23m cubic metres spread over an area of more than five square miles. Into such a maw, the bulk of the UK's hazardous waste could disappear like peanuts into the gut of a whale. Even the small fraction that Minosus wanted to use – 2m cubic metres – was huge, equivalent to sixty football pitches stacked to a height of seven metres. Old salt mines in other European countries had been put to similar use, and here was a rare opportunity for the UK to follow the good example of the Germans and the French. Local activists inevitably saw things rather differently. To them the mine was less of a large, benign whale than a great white shark slavering with malicious intent. Through their neighbourhood en route to its jaws would pass lorryload after lorryload of hazardous waste – stuff that earned its name only by reason of being a 'hazard'. There would be danger on the roads. There would be dust. There would be fumes. There would be noise and pollution. The subsequent public inquiry received more than 2,000 letters of objection.

They were not enough. In July 2002 John Prescott approved his inspector's recommendation and gave Minosus the green light – or thought he did. It is the Fred Karno element in the ODPM's make-up that makes it so endearing: if there is something to be tripped over, then over it will go. In this case it went headlong over the wording of its planning consent – instead of granting permission just for the part Minosus had applied to use, it implicitly allowed waste to be dumped throughout the entire mine. This did not so much open a crack in the legal door as leave it swinging on broken hinges. A local resident duly swept through with an appeal to the High Court, which the deputy prime minister managed to head off only by revoking his own permission. It was, of course, a technicality. Prescott invited another round of written submissions, and the local residents' group was briefly hopeful of a second public inquiry, but it ended with a whimper in December 2003 with a straightforward correction and a reissue of the consent. The necessary Pollution Prevention and Control (PPC) permit from the Environment Agency followed in August 2004. A further legal challenge did hold up progress until early December, but the administrative and legal logjams finally cleared and the first consignments of waste were expected to roll into the mine – three years late – in June 2005.

Without the delays, the mine would have been ready in time to offer relief when it was most needed, on and after 16 July 2004. Its intention was, and still is, to receive twenty-five lorryloads a day, amounting to 100,000 tonnes a year throughout the twenty-year duration of its licence. Much of this will be the dangerously contaminated fly-ash from incinerator chimneys, sealed in one-tonne polypropylene bags, amounting to possibly 70 per cent to 80 per cent of the entire national output. This is exactly the kind of new, breakthrough development Defra had been asking for, and for which the disposal industry itself had been praying. A solution to the fly-ash problem was near the top of everyone's Christmas list. As one department of government looked to relieve its crisis, however, so another had stuck its boot in the door and frustrated it. Thanks to foot-dragging and (to put it kindly) legal misfortune in the Office of the Deputy Prime Minister, the fly-ash has had to wait another year.

Oddly, it was not an issue that Friends of the Earth took an interest in, or was even aware of, and yet – even without the legal sideshow – the salt mine saga neatly encapsulates all the sags and bulges of the waste debate. Rubbish tips are a bit like roads. The more you build, the more traffic you will encourage (which may be a bad thing). But if you don't build enough for the traffic you already have, then you will cause congestion, illegal short-cuts and a rise in the accident rate (which may be a worse thing). And if you block off most of the routes, as the Landfill Directive did in July 2004, it is hard to make much of an argument against exploiting whatever opportunity fortune then provides. We might not like it; we might think waste minimization a more world-friendly policy than waste accumulation; but the hard, like-it-or-not reality is that the flow of hazardous waste is unlikely to abate, that it cannot be wished out of existence and that it has to go somewhere. Every tonne sent to a salt mine is a tonne spared from surface landfill.

The PPC permit allows Minosus, which is co-owned by Salt

Union, operators of the mine, and the waste management company Onyx, to store solid, dry, non-biodegradable, non-radioactive, non-flammable and non-explosive hazardous wastes in sealed containers 170 metres below ground, in conditions which, the Environment Agency says, have been rigorously assessed against the risks of just about everything short of Armageddon – fire, flood, earthquake, even glaciation in case global warming in the next 50,000 years should do a sudden U-turn. The main 'hazard' in the fly-ash and other wastes, it says, will be heavy metals such as lead, mercury and cadmium, which is all the more reason for putting them there: 'The salt mine environment is better suited to the disposal of this type of waste than surface landfill sites, as the salt rock geology of the mine is extremely stable and has remained so since the formation of the salt beds 200m years ago. Storing the waste in the salt mine means that it is isolated from people, surface water and groundwater.'

So that's all right, then? Perhaps it is, given the pragmatic and imperfect world we have to live in. Maybe the least-worst scenario is the best we can hope for. In a face-off between surface landfill and a deep salt mine, the deep salt mine wins. But no space is infinite. We learned that, the hard way, from the oceans. And Winsford Rock Salt Mine, for all its vasty deeps, is no ocean. Even if its licence is extended after twenty years, in time it will become exhausted. It is a good, solid, apparently leakproof piccc of sticking plaster, but sticking plaster is all it can be, a palliative not a cure. In other parts of the country the search must go on for new places to burn or bury waste. It is allowable only because, in the circumstances in which we have placed ourselves, we have no alternative. But let us not be fooled as the waste lorry disappears around the corner. Out of sight is not out of existence. The essential quest is for patterns of manufacture and consumption that weigh less and less on the waste tip, and more and more on the recycling plant. Any system that relies for its perpetuation on the infinite storage of an infinite quantity of hazardous waste is a system abandoned to folly.

Which brings us, neatly and nastily, to the nuclear question. It is, I must confess, a subject of such hideous complexity and of such inflexibly polarized opinion that the strongest temptation is to tiptoe quietly into the next chapter and leave the whole thing to another writer in another book. This is, primarily, a book about waste – how to reduce it, how to get rid of it. In the case of radioactive waste, the questions are rather different: less to do with the treatment of waste than with whether we should create it in the first place. In the twenty-four-hour clock of nuclear history, we are scarcely beyond the first few seconds; and yet there is nothing we can do to halt the ticking. With some wastes likely to remain hazardous for a million years, we are already in it for the long term. Ironically, it is a fact that crudely underpins the pro-nuclear argument. In for a penny, in for a pound. If we have to store nuclear waste for so long anyway, what difference would a bit more make? What are slippery slopes for, if not for sliding down?

The nuclear story in the UK began in the 1940s with the production of plutonium at Windscale, in west Cumbria, for the defence industry, but the really big moment came in October 1956 when the Queen opened the Calder Hall nuclear reactor. Its principal product was still plutonium for nuclear weapons, but it also produced commercial electricity – the first nuclear reactor to do so anywhere in the world. It was with particular satisfaction, therefore, that in March 2003 Friends of the Earth celebrated the closure of this infamous icon of the nuclear age. As part of the sprawling Sellafield (formerly Windscale) complex, it seemed perfectly to symbolize the false dawn and early dusk of the world's most controversial industry. It was, in large part, the events of that early dawn, when radioactive waste went straight into the sea (a practice that has yet to stop), that did as much as anything to consolidate the UK's growing reputation as the dirty man of Europe and to stiffen the opposition. The hyperbole burned away like dew on a furnace. There would be no more talk of 'energy too cheap to meter'. Nine more Magnox-type reactors and fourteen advanced gas-cooled reactors followed Calder Hall, but by 1976 the lights had turned to amber. Under the chairmanship of Sir Brian Flowers, the Royal Commission on Environmental Pollution told the government that there should be no more expansion of the nuclear programme until a safe way had been found to store radioactive waste. The search for that 'safe way' is still going on.

All radioactive waste is not alike. It comes in four degrees of strength – very low level, low level, intermediate level and high level. The very low level stuff (VLLW), mainly from clinical sources (used barium meals and the like), is sufficiently benign to be collected by local authorities and sent to landfill with other, non-radioactive hazardous waste. Low-level material (LLW), though too hazardous for anything but special handling, is still relatively safe. Some of it may not be radioactive at all, being classified only because it has passed through a contaminated area. The great bulk of it is

POISON

contaminated soil, metals or building material, but it also includes, for example, more hazardous hospital waste, and the gloves, overshoes, protective clothing, paper towels, glassware and other equipment worn or used by workers in contact with radioactive material. Shamingly, some low-level radioactive liquids are still pumped into the sea at Sellafield, making the Irish Sea the most radioactively contaminated in the world. Pollution from Sellafield has travelled as far north as the Arctic, and the radionuclide Technetium 99 has been found in the flesh of Norwegian oysters. This hangover from the dirty-old-man years is being phased out, but the abuse of the sea is not scheduled finally to cease until 2020. Most low-level waste does stay on land. It is compacted under high pressure, then placed in metal containers which are filled with concrete before long-term storage in concrete-lined trenches at the national disposal site at Drigg, near Sellafield. Largely because of the contaminated soil and rubble that will come from shut-down and decommissioned nuclear power stations, Drigg will be exhausted by the middle of the twenty-first century, when the UK will have to find either somewhere else to dump its waste or some other way of dealing with it.

Intermediate-level wastes (ILW) are higher again on the hazard scale, though, unlike high-level wastes, they do not generate large amounts of heat as they decay. Most of them are metals (fuel cladding, fuel element debris, pieces of used equipment and other radioactive scrap) and graphite from reactor cores, but there are also sludges from storage ponds and other contaminated liquids. Most are stored in water-filled concrete tanks, or concreted in steel or other specially designed containers, at the sites where they are produced. Two thirds of them are at Sellafield; most of the rest are at nuclear power stations, the UK Atomic Energy Authority (UKAEA) sites at Dounreay and Harwell, and at the Atomic Weapons Establishment at Aldermaston. Other specialized sites receive and store waste from hospitals, universities, research institutes and industrial users of radioactive material. Despite the absence of heat, ILW can be highly radioactive with very long half-lives, requiring isolation for many thousands of years.

High-level wastes (HLW) – mainly spent fuel and the radioactive liquids produced during reprocessing – are the stuff of nightmares, man on the outermost edge of his struggle for dominion over natural forces. This stuff may be contained, but it cannot be controlled. Fatal mistakes are the only kind possible, and the margin for error, not only by ourselves but by uncountable generations over geological time, is nil. The radioactive decay in HLW produces intense heat as well as radiation, greatly complicating the problems of handling and storage. Some 90 per cent of it is at Sellafield, where some of the liquid wastes are vitrified – i.e., incorporated into glass – for long-term cooling in ponds until the heat has reduced sufficiently to allow their permanent disposal, like mythical caged beasts, deep underground. This is the very same problem that so exercised the Flowers committee thirty years ago – to find disposal sites so strong, stable and secure that they could withstand any man-made or natural disaster short of collision with another planet. It is the very same problem that has blown away the 'too cheap to meter' hype and turned the nation's saviour into a national liability. It is a problem that none of the major nuclear-waste-producing countries – the UK, US, Russia and France – so far has managed to crack. Everywhere the search is for 'deep geological disposal' in rock caverns 500–1000 metres below ground, but progress is draggingly slow and only one repository for long-lived, high-level radioactive wastes – the Waste Isolation Pilot Plant (WIPP) in New Mexico – is actually operational (though Sweden and Finland are close to identifying potential sites and the US is eyeing up Yucca Mountain in Nevada). A British company, Amec, meanwhile has devised a version of the vitrification process, which it calls 'geo-melting', that it claims can deliver the permanent solution the world is looking for. It involves heating a mixture of nuclear waste and soil to between 1,300 and 2,000 degrees Centigrade, producing a harder-than-concrete glass block that it says will last for 200,000 years and is so safe that it can be stored on the surface. The US government is building a pilot plant in Washington State, where it will apply the technique to liquid wastes stored after the 1940s atom bomb tests. Will the world be convinced? The technique sounds neat but, however hard the glass and however accurate the calculations, it is not guaranteed to find a place in the hearts of those for whom the surface storage of HLW is Armageddon in a bottle. 'Time will tell', says a sceptical Greenpeace – which, given the geological time-span, at least honours the nuclear tradition of raising more questions than it answers. In the meantime the pent-up monster glowers and grows.

According to the US Institute for Resource and Security Studies in Cambridge, Massachusetts, the nuclear waste at Sellafield is 'one of the world's most dangerous con-centrations of long-lived radioactive material'. For Greenpeace, continued reprocessing at Sellafield would make the UK into 'the biggest nuclear dustbin in the world. Containing massive stockpiles of nuclear weapons-usable plutonium, they are effectively bomb factories and highly vulnerable to terrorist attack.' For or against, massive power generates massive hyperbole. In one sense the story is already over, though in another it has hardly begun. The nuclear power industry as we know it has begun its slow decline into history, its most challenging task now to plot the course of its own demise. No more nuclear power stations are to be built. The remaining Magnox reactor stations will be shut down by 2010, the last gas-cooled reactor station in 2023 and the Sizewell B pressurized water reactor in 2035. Shut down, however, does not mean disposed of. 'Decommissioning' each station – dismantling it, cooling and storing the waste – may take as long as 100 years, and the process is timetabled for completion by our great-grandchildren in 2117. Their own great-grandchildren, and their great-grandchildren's great-grandchildren, and great-great-grandchildren for a thousand millennia, will mount guard on the concrete lid, hostages to the accuracy of our calculations.

The one certainty in the century ahead is that the stockpiles are going to grow. Finding ways of dealing with them is the responsibility of Nirex, a limited company that began life in the 1980s as the Nuclear Industry Radioactive Waste Executive. From its base at Harwell, it monitors nuclear waste in the UK and advises the government, official regulators and waste producers on the safest way to package and store it. The objective (and hard nuts come no harder to crack than this) is to find solutions that are as strong ethically as they are efficient technologically. Many of the ideas in the past have looked more like the products of a Hollywood script conference than serious science – hey! what about sinking high-level waste in deep wells, or hiding it under the polar ice, or leaving it at a 'subduction zone' to be dragged under the earth's crust as one tectonic plate passes beneath another, or dropping it in heavy canisters to bury itself in deep ocean sediments, or blasting it towards the sun? Or maybe it is just skilful propaganda designed to make the burial option, in rock or clay, look attractive by comparison. This certainly is the message you get from Nirex. 'The engineering difficulties and potential for disaster with each of these suggested methods,' it says, 'should be obvious, making it clear why geological disposal seems so attractive an option.'

On 1 April 2001, Nirex reckoned that the radioactive waste already 'in stock' – i.e., stored and awaiting disposal in the UK – amounted to 17,832 tonnes of LLW, 90,371 tonnes of ILW and 3,286 tonnes of HLW. The figure for high-level waste may be a tiny fraction of the whole, but it accounts for well over 90 per cent of the radioactivity. As a rough rule of thumb, 2 per cent of the volume produces 92 per cent of the radiation. The total volume of the 2001 inventory was 92,102 cubic metres of radioactive waste, which Nirex calculates will more than double. As things stand, it predicts future 'arisings' of another 1.66m cubic metres. All of it will – or should be – disposed of in the UK. Nirex takes the view that it is 'not ethical' to export radioactive waste to countries that have had no benefit from it. Controversially, some other countries, including Germany and Japan, still have contracts with British Nuclear Fuels (BNFL) for their spent nuclear fuel to be reprocessed at Sellafield. ('Reprocessing' means separating usable uranium and plutonium from the other waste and returning it, or its equivalent, to its place of origin. In practice this means that the recovered material sent back to Germany or Japan is high in radioactivity but low in volume, leaving the UK to dispose of the high volume of low-activity material that is left behind.)

The resulting shipments of plutonium and high-level waste around the world, their vulnerability to accident or interception, have been one of the most powerful drivers of dissent. Four kilos of plutonium is all it takes to make a bomb. Greenpeace calculates that, in fifteen years, 2,100 kilograms has been shipped from Europe to Japan alone. There is an economic case against reprocessing too. According to the Nuclear Energy Agency of the OECD, reprocessing waste costs at least twice as much as storing it.

It would be very difficult for any sapient human in the United Kingdom to be unaware of the opposition that all this arouses, and few if any local communities can be expected passively to accept any kind of nuclear installation near their homes. Even the transit of radioactive wastes within the UK is enough to cause alarm. A single cask of spent fuel elements, says Greenpeace, 'contains approximately as much radiation as was released in the Chernobyl incident'. The it-could-never-happen, nothing-to-worry-about counter-arguments would carry more weight if Sellafield itself had a better safety record. With the Windscale fire in 1957, it hosted the world's first reported major nuclear accident. Serious radioactive leaks followed in 1973, 1975, 1979, 1981 (when, in an apparent attempt to escape the notoriety of its own past, it changed its name to Sellafield) and 1986. In February 2000, the Health and Safety Executive complained of 'systematic management failure' and shut down part of the plant after quality control records had been falsified. In the same month, HSE began a prosecution against British Nuclear Fuels for the accidental release of concentrated nitric acid inside the plant, an offence for which Carlisle Crown Court imposed a fine of £40,000. Also in 2000, BNFL was prosecuted by both HSE and the

Environment Agency for failing to account for a number of 'sealed radioactive sources' at Sellafield, and in June 2004 the Environment Agency issued an enforcement notice following the company's 'failure to properly maintain' pipelines discharging low-level radioactive waste into the Irish Sea. As with anything to do with nuclear power, there are furious arguments about the health impacts of liquid and gaseous pollution, but it is commonly argued from scientific evidence (a) that Sellafield will cause 200 fatal cancers – the majority of them in Europe and the wider world rather than in the UK – for every year of discharge, and (b) that it will do no such thing. One's preference for one set of lab coats over another therefore is largely a matter of instinct, and it is precisely because few instincts are stronger than self-preservation that the precautionary principle packs such a powerful punch.

Politically, the issue of nuclear waste has long passed the point where it can be embalmed in soothing words, though words there are in plenty. The UK's national policy has been under review since 2001, and has proceeded at the kind of pace you would expect for a million-year strategy. The government is due to publish its decision in 2005, but there will then be a follow-up debate on how the policy should be implemented, with no action on the ground expected before 2007. A new Nuclear Decommissioning Authority eventually will oversee the century-long process of shutting down the nuclear power stations, cleaning the sites and storing the waste. In the meantime, the government in 2004 set up yet another committee, the Committee on Radioactive Waste Management (CoRWM), to which it gave the task not only of finding the best technical solution for nuclear waste but also of selling the idea to the public – a task upon which it was instructed to embark with 'a blank sheet of paper'. In other words, almost thirty years after the Flowers committee first identified the problem, it was being given not just the freedom, but the duty, to re-examine every possibility from the seabed to outer space. This appalled the House of Lords Science and Technology Select Committee, whose report in December 2004 expressed amazement at the breadth of CoRWM's terms of reference, which it found 'dauntingly broad and in some respects astonishingly vague'.

> In the light of the numerous authoritative and exhaustive reports that have been published in the United Kingdom and abroad, of current international agreements, and of European Union guidance, we are astonished that the Committee should have been told to set about this task 'with a blank sheet of paper'. CoRWM could more fruitfully have been instructed to concentrate on the various alternatives for underground repositories that United Kingdom and international opinion have identified as the best options.

If their lordships caught a whiff of procrastination, however, it was not from the standpoint of abolitionists: 'We note that the delay in developing a strategy for handling nuclear waste is seen by the Government as an impediment to considering the role of nuclear power in meeting its objectives of planned reductions in carbon dioxide emissions and a secure energy supply. We deplore this and urge the Government to reconsider.'

On this point at least, the government's caution seems more appropriate than the critics' impatience. In the short term if for no longer, the nuclear argument has been lost on every ground on which it has engaged – environmental, economic, ethical, political. We have done what no generation on the planet has ever done before. Literally on pain of death, for a span of time exponentially greater than the entirety of recorded human history, we have committed our descendants to stewardship of a pent-up force which, were it ever to be released, would cleanse the planet of its most irksome species. You can see how the story would end in the mind of a Hollywood scriptwriter, with a planet of insects.

In the absence of a nuclear option, the power industry in the UK threatens to go into reverse. With coal having been knocked off its perch by the miners' strike of 1984, the power stations have increasingly depended on gas. Since the '70s coal has seen its market share plummet from 70 per cent to 35 per cent, with nuclear in the short term holding at around 22 per cent and gas at 40 per cent and rising. By 2020 there could be no coal-fired power stations left, with only one nuclear scheduled to remain after 2023. The likelihood is that 80 per cent will be gas-fired, with the remainder flying on a wing and a prayer, chasing the 'aspiration' of 20 per cent from wind, sun and tide. Why, then, do we find the chairman of the Institution of Civil Engineers' (ICE) energy board, David Anderson, arguing that the coal-fired stations should not be demolished but kept in working order?

'Our concern,' he says, 'is simple. Britain historically has been self-sufficient in fuel for generating electricity. We have used our own coal, and our own gas from the North Sea. But that's all coming to an end. By 2020 we'll be importing 90 per cent of our gas. Some of it will come from Norway, but the bulk will be from the old Russian republics, Algeria, Nigeria and Iran.

'We are disadvantaged geographically. Western Europe is also investing heavily in gas. The continental grid will be extended, but we'll be right at the end of the pipelines, beyond Germany and France. The further the gas has to travel, the more it will cost.' The disadvantages of remoteness are not just economic. Politically, too, the risks are enormous. Terrorists could shut off the supply, and Britain – indeed, the whole of Europe – could be held to ransom by producer countries' becoming a gas-producing equivalent of OPEC.

Anderson himself does not say so, but one suspects that talk like this may be part of a softening-up process – that the real agenda, explicitly endorsed by ICE and hinted at by the House of Lords Science and Technology Committee, is a new nuclear energy programme. In September 2004, energy strategists at the Department of Trade and Industry told ministers that there was only one way for the UK to meet its targets for reducing greenhouse gas emissions – by producing half its electricity from nuclear power, which, unlike coal- and gas-fired stations, does not emit carbon dioxide. It persuaded *The Times* to lead its front page with the headline 'Britain must go nuclear to help save the planet'. By 2050, it was being said, the country would need forty-five new nuclear reactors. The idea that the shortfall could be met by wind-power was 'wishful thinking', and *The Times* reminded us that the government's Energy White Paper of 2003 had not actually bolted the door on nuclear. The policy it put in place was 'not now but not never'.

It is the kind of talk that makes environmentalists glow in the dark. The worst kind of frustration is to see their own arguments on climate change, for so long opposed by oil-fired conservatives, suddenly seized and twisted against them. Suddenly, instead of mocking the prophets of doom, the oilmen join their ranks. George W. Bush, or whoever it is who puts the words in his mouth, says climate change is an issue. Tony Blair says it is the gravest challenge the planet faces, and that he will force it to the very top of the international political agenda. Instead of praising environmentalists for their prescience, however, the neo-cons wrest the stick from their hands and clobber them with it. 'You cannot remove [nuclear power] from the agenda if you are serious about climate change,' said Blair in July 2004. In the following month, three weeks before *The Times* went nuclear, the *Observer*'s business section skipped off down the same indigo pathway with an article by Blair's former energy minister Brian Wilson. There was less equivocation here than you would get in a piece about eating babies. 'Face the facts', said the headline. 'The future is nuclear'. The first fact we were invited to face was that the homeless victims of the Boscastle flood, having experienced at first hand the savagery of changing weather patterns, should have been 'invited to form views on whether it really makes sense to abandon our only existing significant source of carbon-free energy'. The inference was that they would be gung-ho for a new reactor on Tintagel Head.

Wilson's assertions piled up, one upon another, like herring boxes after a shipwreck. 'In the age of global warming, opposition to nuclear power is a cop-out rather than a rational or responsible position.' How suddenly the wind changes. The one-time radicals who had argued against the promiscuous consumption of fossil fuels (and been proved right) and for massively increased investment in renewable energy (right again) were now to be demonized as reactionaries. If the gas-guzzlers could no longer look forward to a future of promiscuous oil-burning, then we would have to appease them with some other promiscuously available source of power. Anyone who stood in the way – and especially anyone who indulged in the 'fantasy' of wind-power – was an enemy of progress and a stranger to modern ways of thinking. Closed minds had to be unlocked and refreshed; stale ideology flushed away.

'If Tony Benn,' writes Wilson, 'was able to overcome that same ideological barrier almost forty years ago, when left-wing concern about "the nuclear threat" was a great deal more intense, then there is no reason for it to daunt anyone today. But the case, now as then, must be argued. It has to be shown that anti-nuclearism in the age of global warming is a deeply conservative position.'

Wilson does a right-hand, left-hand give-and-take with the deftness of a stealth budget. He calls for 'honest debate', but then warns against actually listening to it: 'The most effective argument against a new nuclear programme is that we still do not know what to do with the waste. But beware those who use this issue as a convenient show-stopper and want to make new-build consecutive upon an "answer" to the question.'

It is beyond the scope of this book to argue the economic case for or against the resumption of nuclear power. That would require a book of its own. Friends of the Earth argues that greenhouse gas emissions can be pared to Kyoto limits by a combination of energy conservation and wind- and wave-power. Proponents of the nuclear industry argue the opposite. Debate is essential, and there can be no objection to the proper weighing of balances. Beware, however, those who use the issue as a convenient show-stopper to invalidate the concerns of those for whom an economic case stripped of ethics is no case at all. For the sake of its credibility, Nirex long campaigned for its ownership to be taken out of the hands of the industry: 'With Nirex still funded and owned by the waste producers,' it argued, 'a public perception exists that long-term radioactive waste management is determined by the financial constraints of the commercial sector rather than the need for safety.'

This perception needs not only to be altered but conclusively to be proved wrong. In July 2004, the government announced that Nirex finally would get its way: the shareholdings of BNFL, British Energy and the others would pass into the ownership of a new company limited by guarantee. We can only hope that Nirex's influence will be as steady as its voice. 'We believe,' it declares, 'that we have a moral responsibility to future generations to find a safe, publicly acceptable and environmentally sustainable long-term solution for dealing with nuclear waste. We cannot and should not expect future generations to clean up after us.'

Not even politicians could find fault with that. Could they?

CHAPTER SEVEN

Scaling the Mountain

AN ADVERTISEMENT PUBLISHED BY THE MINISTRY OF FUEL and Power in 1944 shows 'Mrs Brown' on her knees before a fashionable tiled fireplace, cleaning out her grate. The accompanying text is in the form of a conversation between 'Mrs Brown' and a slightly sinister, silhouetted masculine figure, 'Fuel Watcher', who stands behind her.

'Surely you are not going to throw away those cinders, Mrs Brown!'

'Oh no – I have always found them better than coal for lighting the fire.'

'It isn't only that – it's also a question of saving burnable fuel. Do you know that if cinders were properly used in every household, the home coal consumption of the country would be 750,000 tons less each year?' (A footnote reminds us that this is enough to fire the steel for ten battleships.) The copywriting may not be slick, but the message is as plain as Woolton pie.

The Mrs Browns of the 1940s were expected to do a lot

more than recycle their cinders. The Board of Trade's famous 'Make-do and Mend' poster gives a hint of what was required. A chortling male figure constructed of highly polished pots and pans (colander head, saucepan torso, cheese-grater legs) gambols merrily with a female partner made of thimbles, buttons and bobbins. On a knitting-needle placard she displays a rhyming couplet:

> *Mend and make-do*
> *To save buying new*

Saucepans, frying pans and kettles would be kept bright by an abrasive rub of crushed eggshells or wood-ash, applied with old stockings. Another Board of Trade advertisement from 1943 invites women to brush up their survival skills at a 'Make-do and mend class'. ('Sewing and household jobbery classes and mending parties are being formed all over the country . . .') The same advertisement offers free advice on how to win *New life for old sheets*: 'Watch for signs of wear and deal with a sheet that needs it before there's a hole. Tear or cut it in half lengthwise and join the selvedges in a flat seam by hand. Then machine-hem the outer edges. These thin parts will go under the mattress where there's little strain on them, so your re-made sheet is almost as strong as when new.' There is guidance, too, on 'Things you can turn into towels'.

Nothing was wasted in the kitchen either. Vegetable water just needed an Oxo cube or dab of Bovril to turn itself into soup; breadcrumbs crisped up into breakfast cereal, and potatoes did their bit for the war effort by keeping their skins on.

RUBBISH!

Those who have the will to win,
Cook potatoes in their skin,
Knowing that the sight of peelings,
Deeply hurts Lord Woolton's feelings.

Lord Woolton, after whom the fabled meatless pie was named, was minister for food, in charge of wartime rationing and a home-front general in the war against waste. To prematurely curtail the life of any usable scrap of any useful substance, edible or otherwise, was an act of treachery like cowardice at the front or a refusal to obey orders. Manufacturers of household goods soon learned to trade on their customers' patriotism. 'Your country's fuel supply is vital to the war effort,' declared the makers of Parozone bleach. 'Don't use it unnecessarily! Start your economy on washday by eliminating the need for clothes boiling – use PAROZONE.' Teeth could be mobilized too – by using Kolynos toothpaste you could directly arm the troops. 'Used tubes are wanted for munitions,' went the rubric on the tube. 'Return to chemist.' Gibbs Dentifrice went one better and offered solid refills. 'The patriotic way to buy your Gibbs Dentifrice is in its new battledress refill. This saves the country's metal. Give your teeth a shine with Gibbs in refill form. Popular Size 7d, including tax.'

No one was allowed to forget where the war effort began – right there in the home. Housewives were commanded to save tins for conversion into tanks and aircraft, table scraps for pigswill, boiled bones to make glycerine for explosives, and paper for shell cases. Paper was the big one. *Good Housekeeping* magazine urged its readers to treat it with something not far short of reverence. 'Every time you take your own wrapping paper or carry your purchases unwrapped you are aiding the war effort.' The editor of *Everywoman* went further: 'Any woman who destroys a scrap of precious paper destroys the means of making British weapons just as surely as if she helped to blow up a munitions dump.'

Good Housekeeping fired back with an item headed 'Rubber Robbery': 'You've seen in your papers that the Axis has taken 90 per cent of the world's rubber. Have you done anything about it yet? You may not be able to free the rubber plantations of Malaya and the Dutch East Indies of the enemy, but you can search your home from roof to cellar for Rubber Salvage.

'What about that rubber girdle, that bathing cap, those old tennis balls, that torn bathroom mat? They're wanted NOW.'

And so on. To conserve, reuse and recycle was not so much a question of home economics as of national survival, every bit as important as Digging for Victory or avoiding Careless Whispers. Today, distanced from the age of austerity by the ages of indulgence and infinite expectation, such selfless acts of citizenship, such artless appeals to loyalty, can seem as quaint as cribbage, candle-lit family sing-songs and hand-me-down trousers. What once might have looked like privilege – whatever we want, whenever we want it – we now expect as of right. We are driven by aspiration, and 'prudence' is a joke-word heard only from a son-of-the-manse chancellor of the exchequer. Global warming and the European Landfill Directive are not Hitler and Mussolini; Margaret Beckett is not Winston Churchill; the Environment Agency is not Fighter Command. We do not feel ourselves to be in any way united, or shoulder to shoulder against a common foe. Although we hear the rhetoric, we see little worse than a phoney environmental war urged upon us by vegetarian killjoys who think second helpings are a sin. Sorting our garbage into separate bins is an irksome nuisance, not a willing contribution to the common good. Any woman who destroys a scrap of precious paper is no worse than her neighbour, and fully in tune with the example of her national leaders. It was not Mrs Brown who conjured the fridge mountain; it was the UK government.

For connoisseurs of black humour, if for nobody else, the refrigerator saga was a cracker. Words like 'debacle' and 'fiasco' that were blown about in its wake seem hardly adequate to describe it. Until the summer of 2000, getting rid of an old fridge was as easy as picking up the telephone. You ordered a new one, and the dealer would take away the old to make way for it. It was part of the deal. About half the nation's old fridges followed this route, with the rest being collected by local authorities. For a while it had looked like a model of environmental good practice, a classic circle of virtue. From the dealers' depots, the old fridges were collected by contractors who refurbished and exported them for sale, mainly to West Africa. Some 40 per cent of the UK's used refrigerators found new lives abroad, and another 15 per cent were sold second-hand in this country. There were five distinct benefits. British consumers could dispose of their old fridges without fuss or charge; dealers had a customer-friendly service to offer; people in developing countries had an affordable source of workable second-hand fridges; the contractors made a profit; the profit enabled other, less easily resaleable 'white goods' to be taken away too.

But there was also one large and overwhelming disbenefit. Fridges made before 1996 contained chlorofluorocarbon gases (CFCs) which, when released into the atmosphere, would damage the ozone layer and accelerate global warming. For this reason, the EU banned their export to countries outside its own borders. This killed the export trade, put the contractors out of business and guillotined a useful disposal route for white goods in general. In the ensuing vacuum, local authorities suddenly found themselves having to collect the lot. This might have been OK, except that they were no longer allowed to send fridges to landfill, and after January 2002 they were required to recover 'controlled substances' (i.e., CFCs) from all fridges that contained them. This would need specialist recycling equipment that (who, now, could be surprised?) the UK did not have.

We all know what happened next. With between 2.5 and 3m fridges being discarded every year in the UK, and with existing treatment capacity for just 8,000 a week, municipal waste depots were overwhelmed. In Newport alone, 30,000 fridges were stockpiled within three months. With storage costs escalating, and councils complaining about having to cope with a problem not of their own making, local authorities could no longer afford to collect old fridges free of charge. Faced with having to pay for a historically free service, many people took the only cost-free route still open to them. Fly-tipping soared.

How could it have happened? Same old story. The UK government had known since 1998 that new EU rules were on the way. Instead of mobilizing the waste disposal industry to prepare for them, however, it mobilized civil servants and lawyers to textually deconstruct the wording of Regulation (EC) No. 2037/2000, which covers substances likely to damage the ozone layer, and to challenge its interpretation. To be fair, not all the blame attaches to the UK. The European Commission, too, played its part in sowing confusion. The then environment minister Michael Meacher complained that his department had raised questions of interpretation with the EC on nine separate occasions between February 1999 and June 2001, and had received no satisfactory answer. The argument hinged on whether the regulation required the extraction of CFCs just from the coolant gases in fridge motors, or whether it included the insulation foam too. By the time it was resolved, with an affirmative answer, in the early summer of 2001, it was all too late. The country could not be ready by January 2002.

The government's complaint about European bureaucracy was valid as far as it went, but it didn't go very far. Only two outcomes were possible – either the foam was covered by the regulation or it was not – and a prudent government would have been prepared for both. It could be argued, too, that a government fully committed to environmental protection would have acted out of conviction anyway. The electrical retail and waste management trades both saw clearly enough what was coming. As early as September 2000, Biffa Waste Services wrote to the Department of Trade and Industry, asking it to confirm that foam in domestic refrigerators would be covered by the regulation. With its alert commercial eye, it had spotted both a problem and an opportunity. 'We have been approached by another company interested in developing the necessary technology to degasify these foams and our intention is to undertake a market survey by way of establishing the commercial validity of the system.' To this reasonable and potentially important enquiry, it received no formal reply.

Two months later the electrical goods retailer Dixons raised the alarm in a letter to Michael Meacher. It asked him to confirm that foam would *not* be included. If it was, it said, 'a currently well-managed and environmentally responsible trade will be forced out of existence. This will lead to wholesale indiscriminate dumping which in turn will inevitably result in the release to the atmosphere of CFCs when such products are vandalised or crushed for disposal.'

Biffa had no doubt that the insulating foam *would* be included, and that the question to be asked was not if, but how, it should be handled. In March 2001, it switched its fire to the Environment Agency, with a stark warning about the consequences of inertia – '40,000 tonnes of fridges will inundate civic amenity sites from January next year'. Could the agency define a code of conduct and operating standards for fridge processing equipment in nine months' time, after the regulation came into force in January 2002? As the company explained later to the Environment, Food and Rural Affairs Select Committee: 'Without such assurances from the Agency, we as a company . . . will not invest in the necessary infrastructure since we have no guarantee that others would not undercut us with inferior plant.' To help concentrate the agency's mind, it attached an English translation of the code of practice already adopted by the Germans. Result? 'No recorded reply.'

In June, the bureaucratic mill finally ground out a result – the EC Regulation 2037/2000 Management Committee at last made itself understood to the UK government, and one imponderable at least crystallized into awful certainty. Biffa was right. Domestic fridge foam *would* be covered by the ban; the waste management industry had no infrastructure to deal with it; the problem had been manured by procrastination and now promised to flower into full-blown crisis. What happened next, even by the prevailing standards of hopelessness, was incredible. Nothing did. It would take *four months*, until October 2001, just three months short of the regulation going live, for the government to write to local authorities and warn them of what they faced. Like the fridges themselves, information was being stockpiled rather than processed. It had been like this all the way through. Defra, the department chiefly responsible for negotiating and implementing the regulation, at the crucial moment had not even been aware of the export trade in used fridges which it was about to obliterate – Customs and Excise had not thought to let it know.

For a while, Biffa pressed on. In July it wrote to Michael Meacher, reminding him of the scale of the problem and reiterating its interest in providing the infrastructure to cope with it. In August, with the light dying and the brief hope of summer now sinking inexorably towards a winter of chaos, it wrote again to the Environment Agency: 'We are desperate to make the necessary applications for planning and licensing consent but we can't make a decision on the type specification on the equipment until you tell us what the standards are. Help!'

It also wrote to the Treasury, suggesting various funding options for a national system of fridge recovery, processing and disposal. Letters and emails continued to flutter for a while, but 1 January came and went, the mountain began to rise and, on 14 January, Biffa threw in its hand. It wrote to Meacher withdrawing its offer to invest, and expressing its 'extreme concern' that the absence of clear regulatory standards would once again abandon the trade to the cowboys.

'This letter,' it told the select committee, 'followed from market intelligence to us that a number of companies with no transparent or long-term trading history are now stockpiling fridges in much the same way that others have done for clinical waste and tyres in past years.'

As Biffa had predicted, so it came to pass. In April 2002, Meacher revealed that the cost of processing the backlog might reach £40m – a figure that industry estimates subsequently raised to £80m-plus. Where did all the fridges go? Many went (and still do go) for processing in Germany, whose waste management industry at least had reason to raise a glass to the British way of not quite doing things. Some were legitimately processed in the UK as the local industry slowly cranked itself up. And the rest went ... well, where? Illicit traders do not advertise themselves – they either hide behind legal fronts, or they move in the twilight world of unmarked vans, roadside dumps and sudden conflagrations. We can only know how many offenders the Environment Agency catches and prosecutes: the rest are as invisible as rats in a barn. You know they're there, but you seldom see them. The relationship of the agency with the cowboys continues to echo that of a sheriff's posse with Wild West bad-hats, having more hills and gulches to scour than there are days in the year to scour them. As late as December 2004, there still remained 120,000 fridges to be cleared from five bankrupt storage sites in Greater Manchester – a four-month, £850,000 clean-up operation that required the involvement of the Environment Agency, landowners, Defra, local MPs, Greater Manchester Waste Disposal Authority, the fire brigade and two borough councils.

*

Even as the fridges subsided, an even deeper seismic turbulence was beginning to threaten the landscape. Two more crucial bits of European regulation were on the way – the Waste Electrical and Electronic Equipment (WEEE) and End-of-Life Vehicles (ELV) directives, which will do for electric toothbrushes, televisions, toasters and cars what Regulation 2037/2000 did for fridges. As always, there are deadlines. By 13 August 2005, anyone with a redundant piece of domestic electrical equipment – household appliances, computers, tools, toys – must be able to return it free of charge to a collection point. 'Free of charge', however, does not mean the government is offering to pay. Manufacturers, distributors and retailers must meet the cost of collecting, treating, recycling or disposing of it all; and for products sold after 13 August they will have to accept responsibility for commercial (i.e., non-household) WEEE too. When they sell an appliance they must accept in exchange, 'on a one-to-one basis' and irrespective of make, an equivalent item of WEEE – i.e., toaster for toaster, microwave for microwave. By 2008 they must finance the collection, treatment and 'environmentally sound' disposal of WEEE from people's homes.

On top of this comes what the Environment Agency describes as a 'series of demanding recycling and recovery targets for different categories of appliance'. By December 2006, the industry on average must recover at least 80 per cent of the weight of large household appliances, 60 per cent of small (including tools and toys), and 75 per cent of IT, telecommunications equipment and anything that contains a cathode ray tube. There is even a national target for the annual collection rate – an average of four kilograms of WEEE for each person, collected separately from all other forms of waste. All these targets are likely to increase by 2008.

The ELV directive is very similar. Like electrical/electronic equipment, motor vehicles must be more environmentally friendly at the junkyard stage. Targets have to be set for reusing and recycling vehicles and their component parts. Vehicles themselves must be designed to be more easily re-cyclable, with fewer 'hazardous substances' in them. They are to be broken up only by licensed contractors in authorized treatment facilities, and from 2007 the responsibility for dis-posing of them will pass from the last registered owner to the manufacturer or owner of the 'marque'. By any standards these are radical changes, requiring radical responses from those responsible for imposing, enacting and monitoring them.

Not inappropriately, WEEE is pronounced exactly as it is spelt – as in *wheee!*, here we go again! As Peter Jones puts it: 'We're back to fridges again, aren't we?' Whether or not this bleak prediction turns out to be justified, it is certainly a risk. The government is either calculating with the split-second finesse of an expert juggler, aiming to have a full set of balls in hand by the target date and not a moment sooner; or it has simply chucked everything in the air and has no idea of when or how it will come down again. Nerves are certainly being jangled. One of the loudest yelps was heard in July 2003 from the Better Regulation Task Force – an independent red-tape monitor based at the Cabinet Office which has the impeccably New Labour function of 'advising the government on action to ensure that regulation and its enforcement accord with the five Principles of Good Regulation'. (The five Principles, just to get the full flavour of it, are Proportionality, Accountability, Consistency, Transparency and Targeting – not yet acronymized to PACTT, but there's still time.) Assuming the task force wouldn't tell someone to do something if they were already doing it, the diagnostic tick-boxes all indicated a recurrence of the same old tiredness and lassitude, with the same old injunction to the patient to pull himself together.

In its report 'Environmental Regulation: getting the message across', the task force urged the government to inject a bit of clarity into its policies for ELV and WEEE. 'There is still an opportunity,' said the team leader Janet Russell, 'that the government can avoid another fridges fiasco . . . [It] needs to be clear about what it is trying to achieve; who is responsible for what; being open about when decisions will be made; ensure the waste industry knows the standards it needs to work to; and take a fresh look at how it communicates with its customers. Better project management skills will help.' In the circumstances you wouldn't expect it to be long before that weary old policeman, the Environment, Food and Rural Affairs Select Committee, laid a hand on the miscreants' shoulder. It started to take evidence in November of the same year (2003) and published its report in February 2004 – a speed which, in the turgid pools of Defra and the DTI, must have looked like the Severn Bore. It was certainly flowing through familiar channels and washing up against familiar obstructions. 'Defra,' it said, 'views the negotiation and implementation of Environmental Directives as a painful chore rather than as a positive opportunity for change . . . We again question whether Defra possesses enough of the specialist skills it needs to deal with this sort of Directive, particularly in terms of legal capacity.'

All the same old anchors were dragging. Yet again it had to be explained to the captain on the bridge that no destination could be reached unless someone agreed a timetable and set a course. The manufacturing, retail and waste industries all needed to prepare for August 2005, and would have to invest substantially in organization and infrastructure. For this to happen, they needed to know with absolute certainty what was expected of them. With no clear direction, and with deadlines approaching, 'companies had not been able to reach investment decisions, with the result that there was a risk that there would simply not be systems and equipment in place . . .'. The British Metals Recycling Association said only one of its 2,500 members had felt able to upgrade to the standard likely to be required by the ELV directive. Once again the select committee had to address departments of state as if they were freshers on a business management course: 'The timing of business investment decisions is crucially dependent on having sufficient information available to reach a judgement on when to invest. Invest too soon and there is a risk that new facilities will stand idle; invest too late and business opportunities will be lost.'

If the government failed, said the MPs, there would be another 'fiasco', with increased fly-tipping and abandoned vehicles. Already some 340,000 were being abandoned every year, and West Sussex County Council reported an abandonment rate for redundant vehicles of one in six. Witnesses at select committee hearings competed to see who could come up with the most depressing scenario. With dismantlers likely to charge between £50 and £150 per vehicle, the British Metals Recycling Association reckoned the rate of abandonment before 2007 would quadruple. The British Vehicle Salvage Federation guessed it would rise to 500,000 a year. The Environment Agency predicted only that the number of abandoned vehicles would go up. Much would depend on the volatile price of scrap metal. When the scrap value of a car exceeds the cost of collecting and disassembling it, then the market, both legitimate and otherwise, favours recycling. When the balance tips the other way, the opposite is true.

Questions mounted faster than abandoned Skodas. The British Vehicle Salvage Federation complained of 'confusion and possible disagreement between the various government departments involved . . .' It stood four-square behind – indeed, was largely responsible for – the select committee's own misgivings:

> Whilst we appreciate that Government wishes to avoid a repetition of the fridge fiasco and has extensively consulted as a result, it is our view that the delays, lack of clear advice and instructions has prevented companies from moving forward on investment in appropriate premises, equipment and other facilities to meet the new requirements. This is likely to mean insufficient capacity to deal with ELVs according to the new legislation and therefore a worse situation than the fridge problem.

It was the same with WEEE. How exactly would it be collected and handled? What route would it travel? How could the costs of disposal be apportioned between manufacturers? How could such a system be policed? In practical, hands-on terms, how were manufacturers to dispose of their own used products? The 'producer pays' principle is a cornerstone of all new waste policies, and here was the first opportunity to make it count. The legal framework would have to be tight; the commercial investment massive; the policy comprehensive, definitive and impossible to misunderstand. Otherwise . . .

There would be a shambles. Again, the competence of government departments was called into question. Biffa identified a 'substantial skills gap' in Defra. The British Metals Recycling Association complained of hold-ups caused by a shortage of government lawyers. The British Retail Consortium bemoaned the lack of specialist knowledge and experience among officials. Defra insisted it was comfortably on top of things but the committee, having heard it all before, was not convinced.

> Before embarking on the negotiation of Directives such as the ELV and WEEE Directives the Departments involved should review whether they have available the specialist skills and other resources to do so effectively. The Government should not agree to this sort of European legislation without a clear understanding of *all* of its ramifications ... We strongly recommend that Defra (and other Government Departments such as the Department of Trade and Industry) take steps to develop its understanding of the waste and recycling sector.

The likelihood of misunderstanding was further increased by the number of official fingers in the pie, and by the difficulty anyone outside government (never mind those within) faced when they tried to find out who was supposed to be doing what with whom, and what, if anything, was going on. The DTI, Defra, the Environment Agency and the Driver and Vehicle Licensing Agency (DVLA) all had claims on various aspects of the directives' negotiation, enactment and enforcement, with predictable effects on 'clarity'. Echoing the Better Regulation Task Force, the British Retail Consortium deplored the 'lack of any clear division of responsibility between the DTI and Defra. This confusion has led to unclear channels of communication for stakeholders and often misleading guidance in preparation for implementation.' A representative of the British Vehicle Salvage Federation put it with brutal economy: 'Here we have a bureaucratic nightmare from three joined-up agencies and none of them can make a joined-up decision.'

Attempting clarification, the DTI announced it was to be the 'lead department' in negotiations with the EU. Defra meanwhile would lead on 'certain aspects of domestic implementation' and the Environment Agency in England and Wales (SEPA in Scotland) would deal with enforcement. Leave it to Sir Humphrey. The one thing not in short supply was talk. If the effectiveness of environmental legislation increased in direct proportion to the volume of breath and number of printer cartridges expended, then broken toasters and old Toyotas would convert themselves automatically into wind farms. The DTI itself had set to work in the spring of 2003 with a consultation paper on WEEE, followed by another in November of the same year and another in late July 2004. The deadline for responses to this was 29 October, just 288 days short of the date on which it would all have to be put into practice. There were some big questions left to resolve. How would the 'National Clearing House' proposed for the control and monitoring of WEEE actually work? How would WEEE physically be allocated to its producers? Question after question, right down to whether or not retailers should be allowed to tell their customers how much the disposal of 'historic' WEEE (i.e., stuff sold before 13 August 2005) was costing them, and to charge a so-called 'visible fee' – £1 per appliance had been suggested – to help meet the cost of recycling. It was estimated that more than 10,000 businesses would be affected. Manufacturers, importers, exporters, wholesalers, retailers, repairers, dismantlers, scrap merchants, shredders, reprocessors, landfill operators, even some charities and voluntary organizations – all would need clear instruction on what they were expected to do, and when. All would need to be ready by 13 August. The demands of publishing being what they are (it takes longer to edit, print, publish and distribute a book than it does, apparently, to get a national WEEE network up and running), the consultations were still going on and the crucial decisions still not made when this chapter was written in the early autumn of 2004. Even more companies by now were arguing that there was not enough time left before the August deadline for them to comply with whatever decision the government eventually would get round to making.

One way or another, the WEEE directive in the UK will be the first piece of 'producer responsibility' legislation that makes manufacturers directly responsible for dealing with the waste from their own 'marques'. The ELV directive will follow, though the timetable here is rather longer. Not until 1 January 2007 will car makers be handed responsibility for dismantling and recycling their own products, but an awful lot has to happen before the cogs are meshed. On the operative date, owners of life-expired vehicles must be able to have them collected (or, as the DTI puts it, 'accepted by collection systems') free of charge, even when the cost of processing exceeds the value of the scrap. Processing may be done only at authorized sites meeting specified standards, which must issue official Certificates of Destruction to erase each car, van or lorry from the national vehicle register. Targets for reuse and recycling will increase over time, which means that manufacturers will have to modify their designs to exclude 'hazardous substances', minimize waste, simplify the dismantling process and maximize the recovery and reuse of parts. The current recycling rate for old vehicles is around 80 per cent by weight – a rate which the ELV directive aims to increase to 85 per cent by 2006 and 95 per cent by 2015.

Not the least of the questions being begged is who will actually dispose of the vehicles, and how will the producers pay? The car-makers themselves want to contract directly with dismantlers dealing exclusively with single 'marques' – the end-of-life equivalent of the main dealer system that launches each vehicle on to the road. The British Metals Recycling Association favours a less exclusive approach in which any registered dismantler could deal with any make, and with makers paying indirectly through a central fund. The latter has advantages both economic and practical. The 'proximity principle', which requires waste to be processed as near as possible to its point of origin, is best served by local processors who will accept any vehicle that is offered to them; and it saves processors from having to rely on the patronage of a single client that might at any moment switch its business elsewhere.

One car-breakers' representative suggested the entire system could be funded through taxation – perhaps by a £100 environment tax levied on each new vehicle to cover the cost of an old one being taken off the road. A member of the select committee counter-proposed an increase in the vehicle excise licence. It sounded sensible, workable and cogent but in ideological terms it was fatally flawed. By funnelling money through the public purse, such a scheme would not satisfy the 'producer pays' principle. Never mind that 'producer pays' in effect means 'customer pays', and that the pockets that would be picked for a tax-funded option are the same that would be raided for a privately funded one. It is for similar ideological reasons that the DTI backed the manufacturers' preference for an 'own marque' approach. This was, it argued, 'a pure implementation of that principle that each manufacturer is responsible for the vehicles that they have produced'. We wait to see whether, in this brave new world where destruction becomes a creative act, manufacturers' own-brand junkyards will be as boldly advertised as their showrooms.

Manufacturers, distributors, retailers and trade associations are usually rivalled only by UKIP and the *Sun* in their loathing of all things European. EU directives, and especially environmental ones, spell trouble. New rules mean new expense. Consumers by and large think the same. Tighter regulations mean higher costs; higher costs mean higher prices in the shops. Down with Europe! One issue, however, is different from most of the others, and – if only briefly, like a Christmas truce in the trenches – unites all sides in a common belief. We are all familiar with that right-thinking social phenomenon, the Concerned Shopper.

She is for: muddy carrots, free-range eggs, anything brown.

She is against: e-numbers, factory food, packaging.

Especially packaging. Ostentatiously she refuses the supermarket's plastic bags and wedges everything into a voluminous wicker item of her own. To her, it's common sense. Packaging is unnecessary. It wastes time, resources, money and space in the dustbin. It should be against the law.

As more and more households are asked to separate their garbage and see the 'green' bin brimming with packets, cartons and plastic pots (all nicely rinsed, please), so more and more people come round to her point of view.

Environmental campaign groups have supported her, but manufacturers and retailers have parried with commonsensical thrusts of their own. Packaging is as old as processed food. It goes back to, and far beyond, the early days of selling cheap and piling high. Look: even nature itself packages its products in pod, shell, skin and rind. Well-designed packaging saves energy by taking up less space in lorries and warehouses. It means better hygiene; longer shelf life; easier bulk handling and storage; less waste. For all these reasons it makes food, and every other kind of packaged product, cheaper rather than more expensive. Opposition also carries the risk of killjoy syndrome. Take Easter eggs – extreme examples of over-presentation, in which the product itself becomes part of its own packaging. As a way of serving chocolate it is extravagantly wasteful, impractical and uneconomic. But does anyone actually want to make it illegal? The state needs a pretty good reason to interfere with people's right to spend their money as they choose, and a ban on gift-wrapping would be a gift-wrapped, barn-door target for the anti-Europeans.

And yet the Concerned Shopper is right. Excessive packaging *is* wasteful. Then again, the industry is right too. Good packaging *does* provide all the benefits claimed for it. To bring the two strands together, common sense has to broaden its vision and look for the Goldilocks solution – packaging that is not too much, not too little, but just right. The essential requirements have been summarized thus:

- Packaging shall be so manufactured that the packaging volume and weight be limited to the minimum adequate amount to maintain the necessary level of safety, hygiene and acceptance for the packed product and for the consumer.

- Packaging shall be designed, produced and commercialised in such a way as to permit its ... recovery, including recycling, and to minimise its impact on the environment when packaging wastes or residues from packaging waste management operations are disposed of.

- Packaging shall be so manufactured that the presence of noxious and other hazardous substances and materials . . . is minimised with regard to their presence in emissions, ash or leachate when packaging or residues from management operations or packaging waste are incinerated or landfilled.

As the cod-legal vocabulary may imply, the words are not my own. They are lifted unaltered from the DTI's guidance notes on a revised European directive – the Packaging (Essential Requirements) Regulations 2003 (S.I. 2003 No. 1941) – which came into force on 25 August 2003. The Concerned Shopper may not yet be aware of the benefits, but the trade, almost without precedent in the case of European regulation, has allowed a faint smile to tug at its lips. By passage of law, a tricky internal conflict has been resolved. No longer do environmental, utilitarian and commercial imperatives collide. No longer do companies have to fear that, by cutting back on packaging, they'll be handing the initiative to their more flamboyant competitors. The Packaging (Essential

Requirements) Regulations apply without discrimination. As the DTI puts it: 'The main requirement is that no person who is responsible for packing or filling products into packaging or importing packed or filled packaging into the United Kingdom may place that packaging on the market unless that packaging fulfils the Essential Requirements and is within the Heavy Metal concentration limits [which specify maximum amounts of cadmium, mercury, lead and hexavalent chromium].'

Any member of the public may challenge any item whose packaging they suspect is excessive, and it is open to any trading standards officer to investigate. The threat of unlimited fines, and the certainty of bad publicity, hang over convicted offenders. Given the strength of feeling, it is hardly surprising that many Concerned Shoppers have already pointed the finger, and they have done so to good effect. What it means in practice is that manufacturers are being compelled by law to re-examine their packaging; compelled to use less material and make more of it recoverable; compelled, in effect, to save money. In 2002, even before the revised regulations came into force, refinements to packaging by Nestlé UK resulted in annual savings of 240 tonnes of metal and 208 tonnes of paper and board.

The commitment to 'recycling' offers a good illustration of all the different meanings that much-abused word can bear. Most obviously, a 'certain percentage' must be recoverable in a condition suitable for further use in 'the manufacture of marketable products'. The 'certain percentage' varies according to material. If a package fails on that count, then it must succeed on one of three others. It must submit to 'energy recovery' (i.e., burn well, so that it makes a worthwhile contribution to the heat output from a waste incinerator); or it must make compost; or it must harmlessly biodegrade. All these are hedged about with conditions and qualifications. With materials recycling, the issues are purely practical. The packaging must be designed so that it can be emptied and cleaned sufficiently for it not to interfere with the recycling process. Different materials must be easily identifiable, and separable if they are bound for different processes.

With 'energy recovery', pass marks are awarded if more than half the package weight is organic (wood, wood fibre, cardboard, paper or jute, for example), or thin aluminium foil. Other materials have to pass a 'net calorific gain' test in which the energy needed to ignite the material is measured against the energy released when it burns. In composting and biodegradation, the waste must break down into carbon dioxide, mineral salts, biomass and water or methane. It must disintegrate in the waste treatment process and not reduce the quality of any resulting compost.

Though all this is subject to a blizzard of maths beyond the grasp of most wheelie-bin users, the calculations produce a clear set of results. By the end of 2008, an amended European directive, coming into force in August 2005, will require a recycling rate of 60 per cent by weight for glass, paper and board, 50 per cent for metals, 22.5 per cent for plastics and 15 per cent for wood. If it happens, it will help nudge the UK towards its overall recycling/composting targets for domestic garbage – an optimistic 25 per cent by 2005, rising to 33 per cent by 2015.

Businesses have separate targets of their own. In November 2003, Defra announced that the business target for recovered or recycled paper packaging would increase by yearly increments from 65 per cent in 2004 to 70 per cent in 2008. Glass would go up from 49 per cent to 71 per cent, aluminium

26 per cent to 35.5 per cent, steel 52.5 per cent to 61.5 per cent, plastic 21.5 per cent to 23.5 per cent, wood 18 per cent to 21 per cent – a portmanteau increase for all materials of 63 per cent to 70 per cent. Business waste includes bulk packaging used in transporting goods from producers to processors, processors to distributors and distributors to retailers, but also packaging from products used by customers on commercial premises. A used sugar or sauce sachet from a hotel dining room, for example, or a soap wrapper from a bathroom, is business waste, not domestic. It all adds up. In 2003, the weight of paper packaging recovered or recycled by businesses in England and Wales was 2.09m tonnes. This does not mean that the absolute tonnages of recycled material will go on rising – theoretically, the overall reduction in packaging weight should bring them down even as the recycling rate increases.

It is a rare case of regulatory and commercial pressures pulling together. A report commissioned by the DTI in 2003 from the consultants Perchards showed that yoghurt pots were already 60 per cent, and plastic soft drinks bottles 33 per cent, lighter than they had been thirty years earlier. Plastic carrier bags had halved their weight in twenty years; drinks cartons had slimmed by 60 per cent in ten. For these reasons it was impossible to calculate the precise impact of the new regulations separately from pre-existing trends. Some companies did argue that they were already trying their hardest, but – in a forum not well known for its reluctance to complain – there was a striking absence of protest. Perchards quoted one packing firm: 'The company has a long-standing commitment to packaging optimisation, initially as a means of improving cost-effectiveness. Cost savings usually result in less packaging. The Essential Requirements Regulations have provided significant further impetus . . .'

Where there *was* resistance, company environment managers now enjoyed a significant shift in the balance of power. It put them in a much stronger position to negotiate with sales and marketing departments. Of the twenty-two companies canvassed by Perchards, eighteen said they had reduced their packaging as a direct result of the new regulations. There had been other benefits too. Local authorities, whose trading standards officers have to enforce the regulations, had, in the course of discussion and negotiation, learned a lot more about the practicalities and economics of packaging design, and complaints and investigations had gone down as a result. For their part, manufacturers had learned a lot more about the way their packaging was perceived by the public. Best of all, from the enforcers' point of view, the regulations are 'self-policing'. Every link in the supply chain is pressuring the one ahead of it to cut transport packaging. The one chink in the cardboard curtain is that 'producer responsibility' applies only to larger companies – those that produce annually at least 50 tonnes of packaging waste, or that have an annual turnover of £2m or more. But the small fry are exactly that: minnows in the waste swim. It's the big ones that pack the weight. Perhaps at last the consumer can glimpse some benefit from the existence of multinationals who can cut waste right across Europe.

To help them through the elliptical, self-referential maze of raw EU directives, British companies have a 'Responsible Packaging Code of Practice' drawn up for them by INCPEN (the Industry Council for Packaging and the Environment) and endorsed by Defra and the local authorities' central body

LACORS (Local Authorities Co-ordinators of Regulatory Services). INCPEN's membership includes every link in the packaging chain from material suppliers to packaging manufacturers and producers and retailers of packaged goods. The companies on the list can claim an engagement with, and concern for, environmental issues which those absent from it perhaps cannot. It is a list therefore that needs to be read between the lines, with absentees noted as carefully as those who are named. Here it is in full:

Bird's Eye Walls, Booker, Boots, British American Tobacco, Britvic, Cadbury Schweppes, Chep UK, Coca-Cola Great Britain, Colgate-Palmolive, Coors Brewers, Corus, Crown Cork & Seal, Diageo, DS Smith, Diversey Lever, Duracell, Elizabeth Arden, Four Square, Gallaher, Gillette, GlaxoSmithKline, Imperial Tobacco, Halfords, Information Services International, Institute of Packaging, Kraft Foods, Lever-Fabergé UK, LINPAC Containers International, McDonald's Restaurants, Mars Confectionery, Mars Electronics International, Nestlé UK, Next, Pedigree Masterfoods, Procter & Gamble, Rexam, J Sainsbury, SCA Packaging, Sonoco Europe, Tetra Pak, Thomas's Europe, Trebor Bassett, Unilever Bestfoods.

It is improbable that any of these heavy hitters actually needed INCPEN's recipe for the perfect package, though they may have been glad to see it set down for others (much had been made of the need to 'educate' trading standards officers). First and most obviously, a package must be strong enough to contain and protect its contents in 'normal handling' – which doesn't just mean sitting on a shelf waiting for the Concerned

Shopper to buy it. It includes transport, stacking and – given that stock handling is not a trade dominated by men of gentle disposition   a certain amount of buffeting and abrasion. It may also include compression beneath two kilos of King Edwards, an unwashed swede and a large bag of organic wholemeal flour in the Concerned Shopper's wicker basket. If the package has a handle, then that too must stand up to a degree of rough handling.

Just as importantly, it needs what INCPEN calls 'barrier properties'. It must neither leak nor allow the penetration from outside of anything (air, light, moisture) that could damage its contents. The packaging materials themselves must be hygienic and incapable of imparting odour or other contaminants to either their contents or the environment, and they must not corrode from within. Any stopper, screw-cap or other 'closure system' has to work properly for as long as it is likely to be needed; and the pack in its entirety must not degrade during its 'reasonable anticipated lifespan' (though it is welcome to degrade thereafter).

Next comes honesty. Packages should be no larger than they need to be, and definitely not so big that they create false expectations of their contents. 'A significantly smaller net volume concealed within an apparently larger outer dimension is not acceptable.' Empty 'headspace' inside a container must be no more than is necessary to accommodate unavoidable changes in density (as a product 'settles') or volume (as it expands with changes in temperature). Some leeway may be allowed for Easter eggs and other luxury items where the packaging reflects 'the presentational nature of the product', but such frivolity should not be 'excessive'. Trading standards officers and the Concerned Shopper will have a part to play in determining exactly what passes and what doesn't.

None of this, however, guarantees operational efficiency. There is nothing packed in liquid or semi-liquid form that I have not at some time sprayed over myself or my neighbours. Milk, cream, brown sauce, butter, orange juice, fizzy water, Guinness, sugar – after prolonged resistance to fingernails, car keys and teeth, all have let go like sprung booby-traps with sudden, weapons-grade, high-pressure jets. It is another of those doomed compromises, between imperatives in permanent opposition, that make perfect solutions impossible. Ease of opening conflicts with the need to keep the contents secure against leakage or pilfering. Convenience for disabled people conflicts with child-proofing. 'If the opening method is not obvious,' advises INCPEN, 'clear instructions should be provided.' Here one feels they are rather missing the point. One knows which tab to pull: it's what happens next that's the problem. What *should* happen is swift, safe and, where necessary, total expulsion of the contents. 'The process of opening the package and removing the goods should not damage the contents. Dispensing and pouring ... should not result in waste or spillage. After emptying, residues should be minimal.'

This sounds like our old friend common sense, but common sense is often blind to detail. The perfect milk carton would have a rigid spout, not a soggy cardboard lip that splits the flow like a boulder in a waterfall. But rigid spouts would fail handling, waste-minimization and economic tests. You might as well ask a butter-pat to spread your toast as expect a Tetra Pak to perform like a cream jug. Regulation or no regulation, when perfection fights economy, it's not often that perfection wins. We have to be realistic. If the Concerned Shopper wants minimal waste, then the price has to include the occasional wet tablecloth.

(In a moment that combines serendipity with irony, I pause between paragraphs to make a cup of tea. In the kitchen, I find our rural milkman's glass bottle – perfect for controlled pouring – is empty, so I have to use a plastic one from a supermarket. This is squat, robust and squarish, with a sturdy integral handle that permits a firm and steady grip. It pours like a medieval waterspout from a church roof and turns a cup of hot strong English Breakfast into lukewarm cream-of-tea soup.)

The last phase of a package's life has to be as efficient as the first. Official targets for 'recovery' and 'recycling' have not been well explained to the Concerned Shopper, whose concern would probably deepen if they were. At the very least, her preconceptions would have to be reprocessed. 'Recovery' means getting some, *any*, value back from a piece of waste, if only as heat from incineration. As ever, you need to watch the terminology. Until 2006, the government's aim is to recover at least 94 per cent of business packaging by the more refined art of 'recycling', rising to 95 per cent in 2007 – figures that even the Concerned Shopper might think sound pretty good. 'Sound', however, is not the same as 'are'. 'Recycling' implicitly means converting old material into new (though not necessarily the same) material for further use, but the fondly imagined picture of old cartons for new is as fantastical as the reincarnation of Elvis Presley. 'Recycled' often means no more than 'turned into compost', and it is precisely by this means that many apparently high-achieving local authorities pump up their rates. It takes some of the strain off landfill, but fails abjectly to reduce the demand for new materials. To keep from the incinerator as much as possible of what might be used elsewhere, INCPEN recommends that each different material in a package, including different kinds of plastic, should be marked with an identifying symbol, and that inks, adhesives and coatings should be formulated to not interfere with further processing.

One of the companies involved in the Perchard report was J Sainsbury plc. Of all businesses, supermarkets are the ones most likely to offend the Concerned Shopper and to invite the attention of trading standards officers. Sainsbury's, predictably enough, received a number of complaints about over-packaging, and these serve to illustrate both the complexity of the issues and the value of a well-made challenge. For example: a trading standards department complained that the company's own-label cereal packs were too big. The snag for Sainsbury was that it would incur extra cost if its supplier had to stop the production line and change to different pack sizes for each different customer. In the end, the argument was swung in favour of change by simple arithmetic – a smaller pack would mean a material saving of 12 per cent. Better still, other retailers using the same supplier were persuaded to change as well, so that a single challenge of a single product brought about an industry-wide improvement.

But things are not always so straightforward. Another customer complained that a pack of dishwasher tablets contained less than half the number it would have room for if the tablets were laid flat. 'The reality,' Perchards reported, 'is that the tablets are tumbled into a carton, which is vibrated so that the contents settle . . .' The number in the box was the maximum that could be accommodated by this method, which was the best available. The complaint was withdrawn when it was explained that laying the tablets flat into each carton would need 'a small army of packers' instead of a single machine. It left open the question of whether some future generation of machines might be designed to replicate the space- and material-saving precision of human packers. It is implicit in the INCPEN code of practice that this is what should happen. As paragraph 7.2.1, 'Innovation in materials and products', says: 'New materials and new production and filling technologies should continue to be developed to allow packaging to be made more resource-efficient while maintaining its functional integrity.' It would be difficult to argue that a less-than-half-full pack, for whatever reason, represented the height of the packager's art.

There is another problem too. So many different products are on the shelves (Sainsbury alone stocks more than 23,000) that it is not feasible for every single one of them to have an individually tailored package. There has to be sharing, and some will make a better fit than others. When a trading standards officer complained that the packs used for chewable vitamin C tablets were too big, Sainsbury's replied that its supplier used a range of three different containers, all with child-resistant caps, that were used in the factory for a range of between thirty and fifty different tablets of varying shapes and sizes. In these circumstances, it argued, standardized containers were the only kind that made economic sense.

In other cases, apparent discrepancies between container and content may have more to do with false perception than inflated packaging. Sainsbury's salad containers, for example, are like wine bottles in having concave bottoms, or 'punts'. A customer argued that these constituted false bottoms, which made it look as if the pack contained more salad than it really did. To the satisfaction of the local trading standards department, Sainsbury's argued that the opposite was the case. Before filling, the 'punt' was actually pushed outwards, making the container convex rather than concave, bigger rather than smaller, so that 'as much product as possible' could be packed into it. Only after the lid had been fitted was the punt snapped inwards. This held the salad in place, stopped the leaves sagging, and made the pack itself easier to support on a sloping shelf. Was this a sufficient technical justification for an apparent breach of INCPEN's unqualified injunction, that 'Consumer packaging must not be designed to give a false impression of the nature, quantity or quality of the contents'? In this case the trading standards department decided that it was, though the dislocation between question and answer (ethical question; technical answer) hardly makes it a safe precedent for other cases.

It may also happen that virtue goes unrecognized. Another customer complained that some frozen foods were being sold in thicker packaging. What is often forgotten is that the container the customer buys is the last vestige of a multi-layered system of packaging in which the product completes the various stages of its journey, often in diminishing quantities, from processor via distribution centre to the retailer's stock room and finally to the supermarket shelf. In the case of the frozen food range, strengthening the consumer package had entirely removed the need for a protective outer delivery carton and so had saved several thousand tonnes of corrugated board per year. Case dismissed. For Perchards, the message was fundamental. Critics must assess packaging systems as a whole and should not make judgements based solely on what the consumer takes home and throws away. For the Concerned Shopper it means that the great arbiter, common sense, loses a little of its firepower. On the one hand, more can sometimes mean less; on the other, what she sees in the shops may be only the visible tip of a subterranean package mountain. How can she tell? Who will give her a straight answer?

The local trading standards department should, for one. If she thinks a package in any way fails to meet the Packaging (Essential Requirements) Regulations, then she can seek a ruling. If the trading standards officers think she has a point, then they will take it up with the suppliers, who – if they have listened to INCPEN – will have documentation to prove their compliance with the law. Or not.

Let no one imagine minor design changes are not worth pursuing. For example, Boots used to distribute small cosmetic items – mascaras, lipsticks, etc. – in reusable plastic clampacks. For environmentalists, that word 'reusable' has bells on it. The only grail holier than a reusable clampack is no packaging at all. And yet: when 'life-cycle assessments' were made and all the arithmetic done, it turned out that non-reusable polyethylene bags were not only lighter in transit (reducing the average packaging weight of a three-pack from 24g to 1.23g) but actually saved 30 tonnes of plastic a year. One cosmetic range: 30 tonnes saved. Another 28 tonnes of corrugated cardboard was saved by reducing the outer casings and removing cardboard trays from distribution packs of medicinal creams; and 34 tonnes sweated away by modifications to daily-dose tablet blister packs. Every little bit helps? You bet.

And yet . . . the efforts of the Concerned Packager snipping away, gram by gram, at his blister packs can seem as futile as the Shopper's refusal of a brown paper bag. For every small saving they make, someone else adds more to the deluge. Think what falls out of your weekend newspapers: garden catalogues, travel catalogues, clothing catalogues, kitchen-ware catalogues, furniture catalogues, gift catalogues, gadget catalogues, food catalogues, chocolate catalogues, computer catalogues, electrical goods catalogues . . . a never-ending slither of glossified deals that clogs the waste bin. Consider, too, what the postman brings: more of the same, plus car insurance deals, health insurance deals, loan deals, mortgage deals, charity appeals, holiday deals, theatre and gallery deals, carpet deals, buy-a-new-Volvo deals, make-a-fortune deals – page after printed page, envelope after envelope, all postage paid, a leaf-storm of expensive, resource-gobbling, uninvited trash. Wherever there is a gap in your life, someone will stuff a piece of paper through it (in central London you can't walk down the street or leave an Underground station without running a gauntlet of thrusting leafleteers). It's the same on the internet. Self-protection now requires you to log on to your ISP server and delete all the unwanted rubbish before you download your email. Even then, your screen may be obliterated by a slew of pop-ups. And the telephone! Late at night, weekends, it's all the same. And it's not just the poor, bright-voiced sods in call centres dementedly trying to sell you double glazing or a kitchen. Now you get taped messages ('Congratulations!') in American accents trying to convince you that you've won a prize. Almost the worst thing about it is the sound of our own voices rising in querulous, impotent protest, as if we were mad old fools not in touch with the realities of modern life.

The electronic stuff is an irritation. It, too, gobbles resources

– electronic equipment, electricity and all that goes into generating it – but at least you can kill it with a mouse-click. You don't have to fill your dustbin with it. Numbers in the waste business trail so many noughts behind them that you lose all ability to visualize what they might mean as physical entities. But try it this way. In the 2001 census, the population of London was calculated at 7,172,000. Assume for the sake of argument an average individual bodyweight of ten stones, and you have a combined weight of 71,720,000 stones, or 455,447.7 tonnes. If you put them all on a pair of scales, you would still need to add another 1,488,928 people – roughly equivalent to the populations of Birmingham, Lincoln, Tewkesbury, Tunbridge Wells, Slough and Cambridge, plus a middling premiership football crowd – before they would balance the 550,000 tonnes of bumf distributed every year in the UK in the form of direct mail, door-to-door advertising and newspaper 'confetti'.

The direct mail industry, through its 900-member trade organization the Direct Marketing Association (DMA), argues that the justification for this mighty deluge is its popularity. Again there are some big numbers to absorb. According to the DMA's own figures, in 2003 some 5.4 billion direct mail items were sent, of which 4.2 billion went to private households and the rest to businesses. The result is a total UK consumer spend on direct-mail goods and services of £26 billion a year, with the average mail order customer spending £577 – a figure which I confess to find the most astonishing of all. 'As such,' says the DMA, 'it is fair to say that many customers welcome the information and special offers they receive by post.' So: hands up all those who want more junk mail. In case the economic, just-what-the-customer-wants argument doesn't convince you, then there is an altruistic case too. The DMA argues – one presumes correctly – that many charities depend for their very existence on the support they attract through direct mail campaigns. To ban the mail is to starve African babies, and no one's going to vote for that.

But not for nothing does the DTI in its file-headings bracket 'Junk Mail' with 'Direct Mail'. No matter how many consumers respond to what it offers, the original mailing will not dematerialize. What comes in via the letterbox will have to go out again via the dustcart, all 550,000 tonnes of it. So how does 'producer responsibility' apply here? Unlike the manufacturers of bought and solicited goods – cars, toasters, washing machines, televisions – the bombardiers of unsolicited mailshots are spared the inconvenience of take-back. Junk mail reaches the newsprint mill or municipal incinerator by exactly the same route as newspapers and magazines, through local authority recycling schemes. Private enterprise; public cost. How do they get away with it?

The answer lies in a voluntary agreement carved out by the DMA with the DTI. This involves something called the Mailing Preference Service (MPS), which enables junk-resistant individuals to remove themselves from mailing lists. You can register with it on the internet, though not without facing a last-minute sales pitch: 'Direct Mail,' it insists, 'is a convenient way to shop from home . . . to take their time and make good decisions without pressure, and to get the products they want and need – often at less than they'd pay in the shops.' Some of the cancellations, it acknowledges, are brought about by force of circumstance. 'Many of the people who register their preferences with us have suffered a bereavement, and they simply want to stop future commercial direct mailings from being sent to the deceased.' A minority, however, simply don't want the stuff, and 'as responsible professionals, Direct Mail companies don't want to upset any one or waste time sending marketing messages that are not welcome . . . Equally true, they want to be careful not to waste valuable paper and postage. So registering your choice helps them market considerately, ethically and economically.'

Given their professional expertise, one would expect them to advertise MPS with ruthless efficiency – a mailshot to every household in the UK, perhaps, or information printed on all direct mail material? Well . . . not quite. It's more PR than direct advertising, letting awareness seep into the public consciousness rather than rudely poking into people's homes a message that no one has asked to receive. An advertising campaign in 2002 was reckoned to increase 'awareness' of MPS from 8 per cent at the beginning of the year to 35 per cent at the end. This was followed by an 'awareness campaign' launched in October 2003 with the Australian-based environment group Planet Ark, which was founded in 1991 by the tennis player Pat Cash. There was more national and trade press advertising. Leaflets were distributed in Citizens' Advice Bureaux, and 310,000 postcards were 'distributed nationally in coffee shops and bars'. But still there were no mailshots, and no messages in advertisers' bumf. Was it that the DMA lacked faith in the effectiveness of its own specialist medium, or could it be an *excess* of faith that was the inhibitor? Here, with spelling corrected, is the justification the DMA offered the DTI when proposing its (now agreed) voluntary code of practice, or 'Direct Mail & Promotions Producer Responsibility Scheme':

We see no benefit in displaying prominently the details of the MPS on all direct mail material. Such an approach would do irreparable damage to the direct marketing industry. It would compromise any promotional and marketing message and severely reduce the efficiency of direct mail. Many clients would move out of Direct Mail into other direct marketing media, which would have a severe economic impact on the direct mail supply side of the industry, leading to job losses and company closures.

So: by all means tell people about MPS, but not too many and not too loud. You will know whether or not you are among the cognoscenti, though I have to report that my own attempt to find anyone who had heard of MPS, or who knew what it was, ended with a blank sheet. But maybe it doesn't matter. 'There is a common misperception,' said the DMA in its successful pitch to the DTI, 'that once consumers hear about the service they register, this is not the case, research shows that 9 per cent of people who had heard about the service were very likely to register, for the rest receipt of direct mail is welcomed or not seen as a problem. People likely to register were aged between 45 and 64 and of a higher social grade.' By the early autumn of 2004, 1,812,548 names were registered with MPS. The easiest way to join them is to visit the website, http://www.mpsonline.org.uk, or telephone 020 7291 3310.

Waste minimization, therefore, on DMA's evidence is unlikely to make much difference to the height of the mailshot mountain. If 'producer responsibility' is to have any effect, then it will have to be applied at journey's end, with collection, recovery and recycling. But if it's all destined for the municipal dustcart, and has no specific route of its own, what meaningful contribution can the industry possibly make? The answer is that, although it can't offer to reduce the *quantity* of paper, which of course it needs to maintain if not actually to increase in pursuit of its business, it can at least ensure the *quality* of it.

Advertisers, agencies, printers, data bureaux and mailing houses are all enjoined to ensure that their mailshots are overwhelmingly paper-based; that the paper wherever possible is recycled; that it is light in weight and fit for further recycling with minimal use of contaminants, laminates and adhesives; that offcuts from the production process are themselves recycled; and that inks, varnishes, photographic developers and fixes 'should be reclaimed or recycled where possible'. This is what DMA calls a 'total supply chain initiative'. Will it be enough? It is as much a philosophical question as a practical one. Measuring economic gain against environmental cost is the tank-trap in every battle for 'sustainability'. The difference is almost theological, a question of values and beliefs with no common system of measurement. You think one thing or you think the other. If you're George W. Bush, Kyoto makes no sense because it costs money. If you're an economist, concreting a meadow makes better sense than letting flowers grow. If you're a mail order company, any constraint on your freedom to trade is an unwarrantable infringement of the civil liberties won by Magna Carta. If you are prepared to be called a 'conservationist', then all such arguments swing magnetically through 180 degrees and hold without a flicker.

On the face of it, the DMA has done well out of its deal – a voluntary code with no curb on output. In commercial terms its success or failure will be measured by the accountants in the usual way. But in environmental terms? How do we measure that? Inevitably there are targets. No agreement with any government department, voluntary or otherwise, gets signed off without them. For 2005, the target for recovery or recycling is 30 per cent of the 'delivered volume' of direct-mailed paper, rising to 55 per cent in 2009 and 70 per cent in 2013. But how do we know if the targets are met? End-of-life direct mail travels exactly the same route as end-of-life newspapers and magazines. How can we know how much of the stuff was delivered in the first place, let alone what proportion is recovered? The first question is the easier. The DMA, trade bodies and Royal Mail will measure and record the weight of all direct mail and promotional material delivered to UK households. Weights are routinely recorded as a condition of any normal transaction between producer and deliverer, and the collation of these presents no particular problem.

The second question opens a very different can of statistical worms. The DTI glosses over it thus: 'The Paper Federation of Great Britain will measure the weight of direct mail and promotions recycled at newsprint mills.' It goes on: 'The waste will be reduced by a combination of ease of recovery initiatives, recovery friendly materials and an omission of certain contaminants from the waste stream. Ease of recovery should facilitate increased recycling such as recycled paper mills, composting, and energy recovery.' You will note, once again, the shuffling of the lexicon, the twinning of 'recovery' and 'recycling', and the weaselly implication in the first sentence that some effective monitoring system is in place. It is not.

First we have to question how effectively the Paper Federation of Great Britain (or the Paper Sector Body of the Confederation of Paper Industries, as it elected to be known after January 2004) can measure the weight of direct mail passing through its mills. Does it have vast teams of scrutineers unpicking the bales, separating bumf from newsprint and determining which newspaper inserts are editorial (fashion supplements, say), and which are advertising? Does it know which leaflets or catalogues have been pushed through letterboxes and which picked up from supermarkets, travel agents or other sources? No, of course it doesn't, and neither could it be expected to. When asked, its spokesperson volunteered without prompting that the method in practice was 'not very scientific'. Meaning? 'Newsprint mills will carry out spot checks on what comes in.'

It is not the fault of the papermaking industry, whose commitment to recycling is unquestionable, that the government overstates its capacity for omniscience. Not its fault that yet another environmental 'initiative' unravels at first glance. Yet even if the tally were 100 per cent accurate and not a spot-check lottery, it would still not answer the question. It may give us a reasonable estimate of how much waste is being recycled in the popularly understood sense of new paper being made from old. But the targets are for 'recovery', not recycling, which means they include the conversion of paper into compost and 'energy recovery' through incineration. Is anybody unpicking and monitoring that lot? Will we ever know how much direct mail is being recovered? Do the government's targets really mean anything? At last, from among the thorny thickets of unanswerable conflicts, conundrums and contradictions, we find some easy questions. The answers are no, no, and no.

You cannot blame the DMA for negotiating the best possible deal for its members. That's what it's there for. But again one is left with the impression of industry professionals swerving round the civil servants like footballers through traffic cones. The DTI tries to remind us who is meant to be in charge. 'Maintaining the voluntary approach,' it warns, 'is essential if industry is to avoid the imposition of regulatory fines or levies by the government.' But it is a big stick waved in the dark. The voluntary approach may be maintained with exemplary care and concern for the national interest (cynicism apart, we have no reason to expect otherwise). Or it may not. The targets may be met or not met; they may even be exceeded. In classic Rumsfeld-speak, it is a known unknown: we know the question but we can never know the answer. The government, eyeless in its chamber, can only sit and spin.

We live in a gadget-fixated age. Large areas of our brains have been replaced by electronic aids that programme our lives and stand in for the memory cells we no longer need. We can click from TV to video to DVD, one channel to another and back again, without raising more than a finger. Music pours into our ears from tiny gizmos smaller than cigarette packs. Remote handsets can control ovens, central heating boilers, curtains and blinds, garage doors, anything our overcrowded schedules leave us too busy or too exhausted to do by crude, old-fashioned muscle power. Communities are defined not by geographical neighbourhoods but by numbers stored in mobile telephones, or 'bookmarks' in laptops. Beyond the humdrum toil of utility, we have come to admire gadgetry for gadgetry's sake, finding ever more obscure tasks for our little electronic helpers – boys' toys – to amaze our-selves with.

A gadget-fixated age is a battery-driven age. The serial number LR6 is as familiar as our own postcodes, and we buy them by the dozen. The advantages are huge – tiny, portable packages of controlled and potent power. But there is, of course, a cost. Every live battery, even a rechargeable one, in due time will become a dead battery, and dead batteries are particularly tricky bits of garbage. Some of the stuff that comes out of them – mercury, cadmium, lead – is on the nasty side of hellish. Mercury threatens the developing human nervous system; cadmium is carcinogenic; lead poisoning invites a range of increasingly unpleasant symptoms from diarrhoea to convulsions, coma and death. Batteries containing significant amounts of these – currently about 7 per cent of all batteries on the EU market – are already subject to specific controls that require them to carry identifying labels and prohibit them from the general waste stream. Quarantining the most dangerous 7 per cent, however, doesn't necesarily mean it's a good idea for the 93 per cent residue to be thrown out with the potato peelings. In November 2003, the European Commission proposed a new, much tougher directive that would cover *all* batteries 'irrespective of their shape, weight, composition or use'.

Once adopted in national law, which in the UK is expected to happen in 2007, the result will be yet more targets and further convolution of the waste collection system. Like WEEE, it will involve a head count, with an annual collection target of 160 grams of spent portable batteries – approximately equivalent to eight LR6s – per person to be achieved within four years. At 2004 sales levels, this implies a collection rate of 43 per cent and a significant hike from the current average of two grams per head. For nickel cadmium batteries (the rechargeable kind used in power tools, emergency lighting and other heavy duty equipment), the target is 80 per cent. To make it all work, free collection schemes will have to be set up within twelve months, with a minimum 90 per cent of the collected batteries being recycled. Within three years, an average 55 per cent of material by weight will have to be recycled, except in the case of nickel cadmium batteries, for which the target is *all* the cadmium and 75 per cent of the rest. Industrial and vehicle batteries will be banned from landfills and incinerators. Typical Europe: multiple targets, variable time-scales, simple issues made so complex that (a) no ordinary consumer has a hope of grasping them and (b) the scope for argument stretches towards the infinite. Look forward to an action replay of previous rounds in the open-ended desk-tennis tournament with which the continent's civil servants keep each other on their toes. Look forward to claim and counter-claim; to energetic pursuit and world-weary dismissal of worthy/pointless goals. Do not expect enough infrastructure to be in place in time to recycle all the batteries that are due to be collected (the country's only zinc smelter, at Avonmouth, actually shut during a battery recycling trial in neighbouring Bristol). Do not fear that any excess of excitement will interfere with your sleep. Dead batteries are about as sexy as mating blowflies.

There is a huge tonnage of the things, or a negligible amount hardly worth scoffing at, depending on who you listen to. Consultants hired by the DTI calculated that the weight of 'disposable consumer batteries' – the kind that power our torches, cassette players, radios and so forth – would increase from 17,500 tonnes in 2003 to 21,000 tonnes in 2008 (though industry sources insist it is already much higher). The discard rate for rechargeables over the same period is expected to rise from 2,300 to 3,900 tonnes. It certainly sounds a lot. In its pilot scheme, from September 2002 to September 2003, Bristol set itself a target of 10 tonnes, an estimated 5–6 per cent of all batteries sold in the city, and got very excited when it 'smashed' the target with a final total of nearly 12 tonnes – 'more than the weight of a double-decker bus!', crowed the South West of England Regional Development Agency. Double-decker buses, however, only look big when they are put against something smaller. Stand one on the deck of an air-craft carrier and it would look like a zit on a whale; or, as the British Battery Manufacturers Association (BBMA) might point out, like spent batteries against the rest of the nation's domestic garbage – less than 0.001 per cent of the total. Implied question: why worry? Shouldn't we be looking for more effective ways of reducing waste, like cutting MPs' car allowances or blacking out Channel 4?

The answer lies in the nasties. MPs may burn rubber at public expense, and Channel 4 may give us *Big Brother*, but they can't actually give us cancer. Mercury, lead and cadmium may be the worst, but they are not the only pollutants batteries throw at us. 'Other metals used in batteries, such as zinc, copper, manganese, lithium and nickel may also constitute environmental hazards.' So says the European Commission. Why does it matter? 'In case of incineration, the metals . . . con-tribute to the air emissions and pollute incineration residues. When batteries end up in landfills, the metals contribute to the leachate . . .' In 2002, 45 per cent of portable batteries sold in the EU, some 72,155 tonnes, ended up in incinerators or landfill.

Back comes the BBMA. More than 98 per cent of non-rechargeable batteries, it says, now contain no heavy metals.

Despite extensive tests, no one has ever proved that burning or burying batteries causes any harm to man, beast or tadpole. In this it is supported by the UK government, whose official response to a European discussion paper in April 2003 took issue with the Eurocrats, and in particular with their contention that emissions from spent batteries were 'a significant source of environmental damage and risk to human health'. Existing restrictions, it said, meant that mercury (limited in most batteries to a maximum of 0.0005 per cent by weight, or 2 per cent in button cells) was no longer a major risk, and that the other heavy metals – lead and cadmium – were nothing to worry about. It went on: 'One particular targeted risk assessment carried out on cadmium oxide in batteries by Belgium, established that there was no excessive risk associated with disposal of cadmium-containing batteries to landfill or incineration. The Belgian study also found no evidence to support that there is a risk to the environment or human health associated with the disposal of lead-containing batteries.'

The last point at least is pretty much beyond dispute. Lead acid batteries are the heavyweights used in industry and motor vehicles. They do not find their way into the domestic waste stream and the value of the lead ensures that almost all of the metal is recycled. As always, it is the commercial imperatives that speak loudest. Where recycling cuts cost or makes profit, you don't need the force of law to make it happen. Even George W. Bush could get behind an environmental policy that made money. One might suppose, then, that the treasure trove of valuable metals – nickel, cobalt, silver – in old batteries would unleash a kind of Ever-Ready gold rush, with prospectors raiding the dustbins and muggers forcing you to empty your Walkman. The European Commission comes up with some big numbers. The proposed new Battery Directive, it says, every year would save 20,000 tonnes each of manganese and zinc, 15,000 tonnes of iron, 7,500 of lead, 2,000 of nickel, 1,500 of cadmium and 28 of mercury. In terms even of double-decker buses this is impressive – equivalent to some 8,440 72-seater London Routemasters. It looks like a clincher. Recycling and reusing this amount of metal will save resources, save energy and reduce global warming in an endless cycle of renewal – the cherished 'closed loop' that is the *sine qua non* of good practice . . .

(Here again I interrupt myself. The date is 4 November 2004. On BBC Radio Four's *Today* programme, a scientific adviser to the newly re-elected President Bush has been asked about America's attitude to global warming. He is a straight talker in the presidential mould; doesn't mess with fancy phrases; gives it to us straight. The facts, he says, are these. The UK government's chief scientific adviser, Sir David King, is an ignoramus. European climatologists generally are in the pay of governments with a vested interest in damaging the US economy. The whole global warming thing, with its absurdist idea that honest Americans should rein back on their world-leading energy consumption, is an international conspiracy designed to thwart the legitimate progress of God's own superpower. The rightful place for climatologists, we may infer, is in the global slop bucket with gay abortionists and anyone whose religion involves an accumulation of facial hair. In the circumstances, the responsible disposal of my annual LR6 output looks about as useful as a shower cap under Niagara. Worse: by association, it makes me an anti-American evildoer.)

As it happens, conflict with the self-canonizing ideology of a morally crusading White House is not the only problem the Battery Directive faces. There are difficulties in the real world too, though the economic antichrists at the European Commission go on building their case. Using recycled cadmium, they calculate, will require respectively 46 per cent and 75 per cent less 'primary energy' than starting from raw. 'For zinc, the relation between the energy needed for recycling and the energy needed for extraction from primary resources is 2.2 to 8.' Overall, they propose to recycle all the lead, all the cadmium, and 55 per cent of all the rest. Even a right-wing God alert to the needs of his own planet might nudge George W. in the direction of that one.

But of course it's not that simple. Indeed, its usefulness as an example may be precisely this, that it exemplifies the complexities and conflicts that bedevil any argument about recycling. Self-evidently recycling is desirable. Self-evidently it incurs a cost. Self-evidently a balance needs to be struck. It is in adjusting the scales, and in determining the precise angle of tilt, that the rift opens up. As usual, the UK is on the side of the doubters. Armed with its consultants' findings, the DTI does not find the balance to its liking. It warms up with a few easy quibbles (the italics are the department's own):

Different targets are discussed for '*all batteries*', for automotive batteries and for batteries containing cadmium. This proposal is at odds with the introductory rationale that the current Battery Directives are '*a source of confusion for customers*'. A range of targets, and divergent collection initiatives is likely to perpetuate any confusion that exists currently.

The highest targets exceed those attained anywhere in the EU, and, as a result, there must be concern as to whether they can be achieved consistently across the Community.

And then it really gets down to business:

Collection of batteries, for all types, and through whatever means, will have significant environmental and economic impacts. There will be financial costs associated with building householder awareness of the need to separate batteries through various media, with the provision of battery collection bags, bins and boxes, with storage, bulking and transport, and with the administration of the collection system. These operations could also have significant impacts associated with the consumption of materials and energy.

In the consultants' opinion, the financial and environmental cost of all this would be 'only offset to a limited extent' by the savings in materials and energy achieved by recycling. They could see no benefit in reclaiming my LR6s at all. 'The benefits of extending the Directive to consumer batteries are tenuous, since the environmental significance of disposing of the majority of the materials they contain is hard to ascertain.'

The European Commission's big numbers, its 8,000-plus fleet of zinc and manganese Routemasters, looked less like transports of economic delight than an expensive traffic jam. As recovery rates rose, the consultants said, so would the environmental costs of dealing with them. 'If the Directive were restricted to industrial and automotive batteries and lead- and cadmium-containing consumer batteries, it would be easier to justify on the grounds of environmental impact and cost effectiveness.' Presumably for reasons of consistency, having argued that different rules for different batteries would 'perpetuate confusion', the government omitted this point from its official response. But it seized on the economic point. The reason our bins were not being raided by battery prospectors, and the current recycling rate for consumer batteries was so low – the commonest estimate was a mere 2 per cent – was not hard to seek. '[It] is largely because there is insufficient value in the materials contained in the batteries to warrant their being recycled.'

All of which, with resounding echoes of hazardous waste, refrigerators and WEEE, begs the biggest question of them all. It's all very well the government invoking the 'producer pays' principle, and requiring manufacturers and retailers to finance and run take-back schemes, but what then? Recovering and recycling the metal would require specialist treatment plants, and we just don't have them. The reason for running a trial campaign in Bristol, part-funded by the DTI, was the proximity of the UK's only zinc smelter, operated by Britannia Zinc Ltd (BZL) at Avonmouth. At the time, it was believed that BZL, which produced some 100,000 tonnes of refined zinc a year, would have the long-term capacity to receive and recycle the country's entire output of batteries. The prospect of a zinc bonanza, however, did not cause share prices to soar. On the contrary. In October 2002, only a month after the local battery scheme began, the depressed state of the global market persuaded BZL's Australian parent company, MIM Holdings, to cease smelting in Europe. On 24 February 2003, the Avonmouth plant went cold, and with it went the entire existing market for recycled batteries in the UK.

According to the DTI, the most practical disposal/recycling

route for consumer batteries would be through a steelmaker's electric arc furnace. This would permit the recovery of the small amounts of steel they contain, but it would not get anywhere near justifying the cost. 'High recycling rates could only be achieved by using the zinc dust removed as a waste from the electric arc furnace, or through other pyrometallurgical processing.' The demise of BZL meant the dust could not be recycled in the UK and would have to be exported to other countries in the EU.

The financial and environmental costs of this would far outweigh the value of the materials recovered, as well as the environmental benefits of recovering materials from the spent cells. Similarly, there are no other pyrometallurgical plants accepting primary or cadmium-containing batteries in the UK, and considerable work will be needed to develop the necessary infrastructure to recycle batteries in the UK.

If, when and by whom the 'considerable work' would be done was another of the unanswered questions that UK policymakers like to leave swinging in the wind. Insofar as any policy was discernible, it was characterized by negativity. Only problems were envisaged. How could we monitor the recycling rates of batteries passing through processing plants? (Possible answer: take a leaf out of the junk mail book and have a stab in the dark.) How could we calculate annual rates of return when the user-life of a battery might be as long as three years? Should the recycling rate in 2011 be expressed as a percentage of battery sales in 2008? And percentage of what? Overall weight, or numbers sold? And so on. In a way, this could all be seen as the weft and warp of the administrative process – all very necessary, and in the public interest, to rub the edges off rough-hewn legislation and plane away all the bumps and blemishes of impracticability or ambiguity. But it is all so agonizingly slow, to the extent that the process itself becomes the greatest obstacle to its own successful conclusion. In the spring of 2003, the UK government already was raising doubts about its ability to meet a deadline in 2008. Nineteen months later, in November 2004, I asked Defra what progress it had made.

By way of reply, it referred me to a company called G&P Batteries Ltd, which it said was setting up a new processing plant in West Bromwich. This was true. But the company was doing so despite having 'failed spectacularly' to persuade either the government or its appointed recycling agency, the Waste Resources Action Programme (of which more in Chapter 8), to help with funding. It was sticking its neck out, ahead of detailed legislation that would define the standard to which it had to operate, in the hope that it would 'stimulate more collections and get the volume up'. To act alone without legislative certainty was a commercial risk, but one it felt was worth taking. The plant, due to begin operation in January 2005, would separate battery waste into three 'fractions' – paper and plastic, steel casings, and so-called 'black mass', a dark powder containing zinc and manganese. The paper and plastic were waste, destined for landfill, but the steel (obviously) and black mass (potentially) could be sold for recycling. In November 2004 the company was still exploring new markets and applications for black mass, which can be made into new metal, or processed into chemicals, but was hopeful it could find buyers in the UK and would not have to export to Europe. If it succeeds, it will save a rasher or two of the government's bacon. But one brave, pioneering company does not add up to a national policy.

By some estimates, the volume of discarded consumer batteries each year is already close to 30,000 tonnes, of which 80 per cent are of the kind G&P is able to receive – i.e., alkaline manganese (used in personal stereos and radio-cassette players) or zinc-carbon (torches, clocks, shavers, radios). In 2003, G&P processed just 130 tonnes of these. From 2005, if it runs the plant for twenty-four hours a day, it will be able to handle a maximum of 1,500 tonnes. That is 192 London double-deckers, which, at a conservative estimate, leaves just 3,195 looking for a place to go.

CHAPTER EIGHT
Round the Corner

NOVEMBER 2004. ODDLY, AS I STAND ON AN ELEVATED walkway inside a waste separation plant near Norwich, I am reminded of Scottish salmon. The real thing, that is – the truly wild North Atlantic fish, *Salmo salar*, not the flaccid, fatty cage-potato of the supermarket chill cabinet.

This is, I concede, an odd thought with which to begin a description of a waste facility. Not much ocean freshness here. The air is dry with dust, not wet with spray, and the thunder of heavy machinery could hardly be mistaken for the crash of breaking water or the whining of fishermen's reels. And yet the simile is exact. Before me runs a fast-moving stream of paper on a conveyor. Every few seconds an individual piece leaps high into the air, twists and disappears while the stream below surges inexorably onward. The only faults in the analogy are that in this case the 'salmon' are swimming with the flow rather than against it, and that their leaping is a mark not of lusty exuberance but rather of banishment in disgrace. Like the real thing, however, they have got this far only because they escaped the attentions of fishermen downstream. Very much unlike the real thing, the leaps are duplicated as tiny coloured dots flipping across a screen. It is all authentically mesmerizing.

Downriver, the ocean parallel collapses into bathos. The journey begins, without romance, at the head of the Norfolk waste stream, to which I am a minor contributor. Inside my gate since July have stood two large wheelie bins, one grey, one green, each holding 240 litres. The council dustcart empties each one on alternate Tuesdays – grey one week, green the next. It is with a genuine feeling of guilt that I confess I've never been outside early enough to see it in action – the bin-men in rural North Norfolk are beaten for early rising only by the blackbirds and sugar beet lorries – but the evidence all points to speed and efficiency. No litter is spilled in the road or garden; and if, as occasionally happens, I forget to put out the right bin, the men come in and fetch it themselves. Bravo for them.

In return, I do all I can to be a good citizen. In the kitchen are two smaller bins of my own. Into the first of these go junk mail and other clean waste paper, cardboard, drinks cans, foil, rinsed plastic bottles (caps removed) and yoghurt pots (washed and dried). Into the other goes all the rest. The former empties straight into the green wheelie bin. The latter, sealed in plastic bin liners, goes into the grey. Newspapers are bagged separately and left out for collection on green bin days. Bottles and other glass go, unsorted, into the village bottle bank. Other stuff when it arises – textiles, old computer keyboards, voiceless answerphones, chainless bikes – to a civic amenity site just outside Wells-next-the-Sea, whose custodians have developed the collection of rubbish into a new kind of community art form. There are separate skips for every kind of recyclable material from rubble to books, and the brute functionality of the place is sensitively softened with flowers, like a memorial garden for your emptied attic. Bravo for them, too.

But back to the wheelie bins. The bagged-up contents of the grey one still follow in the footsteps of the cave-dweller to a landfill site. The unbagged contents of the green one are taken straight to the county-council-owned (but commercially run) 'materials recovery facility' on a bleak – you could hardly expect it to be otherwise – half-built industrial site at Costessey, not far from the Norfolk Showground on the furthest outskirts of Norwich. A neighbouring landfill 'cell', being fed by dustcarts and levelled by bulldozers, lies beneath a snowfall of trash-crazed gulls that are distracted only briefly by the bang of a bird-scarer. In due course, the car park may disappear beneath an anaerobic digester or some other, more mannerly treatment system that will replace the landfill once the government gets its policy straight. As always, there is time pressure. To design, plan and build a plant on the necessary scale will take around five years, which is the approximate, estimated lifespan of Norfolk's remaining landfills. With time leaching away, every lost day potentially is a day of disaster banked for the future.

Part of the future, however (and for once it is a good part), is already here, though it can't be said that many citizens of the present have gone out of their way to welcome it. This whole business of sorting stuff into separate bins, of having to *think* about rubbish rather than blindly chuck it away, does not strike all council-tax payers as value for money. That the reverse is true – that the value of the service increases proportionately with rising need – is a stark illustration of the government's failure to drive home a message that is every bit as pressing as wartime thrift. 'Margaret Beckett,' said one senior official in a large English city, 'neither understands nor cares.' One problem with recycling has been the lack of any clear consensus, or firm official guidance, on what might constitute 'best practice'. There are targets to shoot at, but not much agreement on how best to aim and fire. Which materials should be recycled? How and when should they be sorted? By the householder into a range of colour-coded bins and boxes? By the dustmen at the kerbside? By hand at a processing plant? Automatically, by machine? Which method produces the most recyclable material? Is this method (whatever it is) inevitably the one that will yield the greatest benefit for taxpayers and make the best economic sense? Is *all* recycling unquestionably good?

There is no single or simple answer. What is right for a rural county like Norfolk, with its thinly dispersed population, might make no sense for a densely populated city. Rural, however, need not mean backward. Norfolk Environmental Waste Services – reduced, inevitably, to NEWS – has invested £6.5m in technology imported from San Diego, California, and it's as close to rock 'n' roll as a waste plant will ever get. It's the first of its kind in Europe, one of only seven in the world. I am led initially to a huge conference room with cinema-size DVD screen and rows of expectant chairs that testify to the stream of visitors it has received since it opened in April 2004. Through a large internal window you can watch the lorries turn into the reception hall, where they disgorge on to the concrete floor great slews of 'dry recyclables' – paper, cans, plastic. If it's a Tuesday, my own stuff is here by 11am. One

JCB shoves the new stuff on to the mountain. Another nibbles away at the base, shovelling it into a pit whence the first of many conveyors will feed it into the plant. Appropriately in a place called NEWS, the dominant background colour is of newsprint, though the stack is sequinned with the many colours of junk mail, glossy magazines, drinks cans, plastic bottles. You know even before you utter it that your joke about the Turner Prize will not be original.

NEWS has a theoretical maximum capacity of 90,000 tonnes a year, processed at a rate of up to 15 tonnes an hour. With six local authorities so far signed up, it is operating profitably at just under half capacity, running each day from six in the morning until ten at night. The carrot for the councils is that the more they deliver for recycling the less they will have to pay at the gate, with the further temptation of a 25 per cent share of profits after the plant has met its processing costs. The morning I am there, it is running at 12 tonnes per hour – 0.6 of a tonne for each of the twenty men working the shift. In the vast, hangar-sized machine hall there is no smell beyond the faint, mushroomy odour of paper, but a great deal of dust and noise. The few humans are like tick birds on the hide of a rhino, insignificant soft-tissued morsels hardly worth swallowing. The machine feeds itself through a gigantic network of speeding belts that, in their criss-crossing, computer-controlled complexity, suggest a coming together of M. C. Escher, William Heath Robinson and NASA. If there's a glitch, computer nerds fix it from San Diego. It's pointless trying not to be impressed.

From the 'pit' at the delivery point, the stuff cascades down a chute on to a rising conveyor that whisks it up and through a 'pre-sort' cabin, in which a small team of sorters picks off all the brown cardboard (which is flung straight down chutes to a baling machine) and items of unrecyclable junk which are removed from the conveyor and diverted to landfill. One is tempted to make comparisons with the Victorian dust-yard – human gulls pecking over the detritus – but the analogy is false. Insofar as it is possible for any waste-handling area to be 'clean', this one is. The cabin is heated, sound-proofed, air-conditioned and pressurized to keep down the dust. Many of the people who work in it have been recruited from the food industry, lured by better conditions and higher pay. The pickers, working fast, don't catch quite all the cardboard, or all the foreign bodies, but they do catch most (the conveyor is deliberately run at high speed to help spread the material out).

Beyond the cabin, the jumble of paper, plastic and metal tumbles down another chute into the technological marvel at the heart of the process, the so-called 'V-screen'. This consists of two gigantic, sloping banks, or 'wings', of rubber-clad heavy steel rollers, nine to each wing, that meet at the bottom to form the V that gives the thing its name. It looks like a Bond-movie man-eater, an indiscriminate muncher between whose whirling rollers anything and everything would be pulped or cut to atoms. The noise alone is terrifying. And yet, for all its thunderous energy, it is a strangely light-fingered giant that can tell the difference between paper and plastic and, with 85 per cent efficiency, sift them on to the right delivery lines. It has the simplicity of genius. Flat, two-dimensional objects – i.e., sheets of paper – are lifted and flipped over the top of the wings by the upwardly rotating rollers, and thence on to the next conveyor. Three-dimensional objects – plastic bottles, food and drinks cans – cannot climb but simply bounce up and down like a cloud of multicoloured fireflies at the bottom. They either fall between the rollers or out through the open end of the V on to a conveyor of their own.

Before we trace their progress through and beyond the plant, we should look at the influence on such a system of public stupidity, eccentricity and malice. Most obviously, it is what makes the pre-sort cabin necessary. The pickers have to remove as much as possible, as early as possible, of the stuff that shouldn't have been there in the first place. The problem public itself is sortable into three broad categories. The more-or-less incorrigible, anti-social couldn't-give-a-tossers who chuck stuff into their black and green bins without discrimination. The honestly confused, who can't understand what goes where, or why it's necessary. And the over-eager recyclers. The last two categories have substantial areas of overlap, caused very often by poor advice handed out by local authorities when they delivered the wheelie bins. Like many others, I had been rinsing and green-binning my plastic yoghurt pots because that is what the council told me to do. But NEWS doesn't want them. They have no value as re-cyclables and have to be discarded. It doesn't want supermarket bags either (take them back to the shop); or film-wrap (including the bags your weekend newspaper supplements now come in); or Christmas wrapping paper (much of it is heavily inked plasticized foil, not paper at all); or metal in any form other than cans or spent aerosols (take anything else to the household waste recycling centre); or aluminium foil; or polystyrene food trays; or waxed cardboard drinks cartons; or plastic in any form other than bottles; or anything organic; or textiles; or bottles; or the envelopes that junk mail is delivered in (the glue is a contaminant). None of these at the moment is economically recyclable through the automated sorting plant. All this is hard enough for the people who live here to remember. For the thousands of holiday visitors to the Broads and Norfolk coasts, who may come from places that have different schemes or no scheme at all, it is beyond comprehension. And for the Concerned Shopper it is yet another affront to common sense. Why don't they want her washed-out margarine tubs? Why can't they *properly* engage with the war on waste and recognize that *everything* has a value? In her secondary role as over-enthusiastic re-cycler, she is a quite magnificently accomplished nuisance. Not only does she attempt to recycle things that can't be recycled, she may even take the trouble to flatten her plastic bottles. This confounds the V-screen, which thinks they are paper.

Like every institution that wrestles with the human psyche, NEWS already has a black gallery of rogue exhibits. Objects picked off the line include a 5lb sledgehammer head, a sickle, a bag of nails, an alarm clock, an air pistol, a live twelve-bore shotgun cartridge and an American M16 rifle clip containing fifteen rounds of live ammunition. A cold water tap that escaped detection blocked a screen and closed the entire plant for an hour. Other goodies served up to the sorters include, inevitably, bags of mixed garbage from the Concerned Shopper's extreme opposite, the couldn't-give-a-tosser, who also like to chuck in disposable nappies, dog faeces and the literally life-threatening scrap which the trade knows as 'medical sharps'.

Racing away from the V-screen now is a fast-moving stream of all kinds of paper and card – mostly newspapers and magazines but also bits of food wrapping, old school essays, shopping lists, junk mail, cereal packets ('grey card' in the jargon), brown paper bags, plastic film, a few bits of brown cardboard that escaped the picking line, the Concerned Shopper's flattened plastic water bottles. The real paydirt is newspapers and magazines. The price fluctuates but typically they sell to a North Wales recycling mill for between £40 and £50 a tonne. The mill wants glossy magazines in the mix – being made from virgin wood pulp and not previously re-cycled paper, they are needed to 'refresh' the stock – but it doesn't want tatty scraps of junk mail or other miscellaneous bits and pieces. These have to be separated out and sold at lower value (typically £25 a tonne) as 'mixed paper' to a mill in Kent that turns them into grey card for packaging. The separation and 'cleaning' processes are two more little peaks on the graph of human genius. First, the 'mixed paper' has to be removed. This is done at the next 'screen' – a series of rotating axles with discs placed at calculated intervals. When the paper stream hits these, the larger, flat sheets flow on over the top while the smaller bits bounce and drop through on to another conveyor beneath. Both streams then hit a clean-up line.

It's so clever, it's almost funny: you just want to stand and watch. Two banks of infra-red lights shine down on to the belt, which by now is moving at six metres per second. The enabling factor is that different materials reflect and absorb infra-red light at different rates, thus giving each one its own (more or less) unique infra-red 'signature'. The computer has been trained rather in the manner of a sniffer dog – being shown various materials, recording their signatures and being taught which ones to let through (in this case clean paper) and which ones to stop (everything else). When the detector spots a bit of plastic film, or brown card, or anything else it doesn't like the look of, the computer fires a jet of air that pings it up out of the stream, over a barrier and down the reject chute. The comparison with a salmon leap is impossible to resist. Coloured bars on a screen identify every fragment as it either passes or is blown away. The only escapees are a few small pieces of plastic whose signature is close to the glossy magazines', and these will have to be picked at the end of the line by hand.

The plastic bottles, steel and aluminium cans meanwhile leave the V-screen on a belt of their own. The steel will be picked off by electromagnet, and the aluminium ones by something called an 'eddy current separator', which uses a rotating magnetic field to levitate and eject non-ferrous metals (I would explain it more fully if I understood it myself). The steel cans are baled, not crushed, and sent to Hartlepool, where they have the tin stripped off, then on to the Midlands or South Wales to be smelted back into new cans. Each bale weighs around a third of a tonne and is worth less than £10. The aluminium cans are crushed into dense, heavy ingots that sell at £700 a tonne. Their cobbled metallic texture and kaleidoscopic colour make them a favourite photographic subject for art students. They go to Birmingham or Warrington, where an oven burns off the lacquer at a temperature of 500 degrees. And this is why NEWS doesn't want aluminium foil, or any other contaminant (paper or plastic) mixed in with them. At these temperatures foil will just flare and dematerialize in a puff of black powder. The problem is that the flaring raises the temperature in the oven, which has a fail-safe mechanism that shuts it down when it overheats. The Concerned Shopper in this context is a species of dangerous, fire-breathing monster.

At some time in the future the Norwich plant might add

equipment to sort plastic bottles into three discrete streams – polyethylene terephthalate (PET), used for fizzy drink or water bottles; high density polyethylene (HDPE), used for plastic milk and juice bottles; and polypropylene (PP), used in heavy-duty bottles for shampoo, washing liquids and household bleach. At the moment it collects and sells only mixed plastic, which goes to markets in the UK, Europe and the Far East, where it is turned into things like garden furniture and plastic drainpipes. It will be the market that decides whether the extra value of separated plastics is sufficient to justify the cost of installing expensive equipment to sort them. And this, really, is the nub of it. In a world where 'sustainability' has become a one-word ideology for almost everything, there is no hope for a product that doesn't turn a profit. At the beginning, NEWS separated and baled up plastic bags (the dedicated chutes are still there in the pre-sort cabin), but the material was so contaminated that it was worthless. Many of the bags that appeared on the line had other stuff inside them, and much of this was not only unrecyclable but disgusting and dangerous. Now they go straight down the waste chute for landfill.

The technical whiz-bangs at the NEWS plant are a bit like the architecture at Poundbury: they may blind you to the real issues, persuade you to believe that the whole world could be like this, or that theirs is the best or only answer. The fact is that there are many other ways of sorting and separating waste; other choices that could be made about which materials to recycle; other methods of collection and transport. One local authority often cited as an exemplar of best practice is Daventry in Northamptonshire. Unlike Norfolk, which currently relies on bottle-banks, it includes glass in its household collection and expects its citizens to make rather more of an effort. The council provides each household with *four* different containers – a red plastic box, a blue plastic box and two wheelie bins, one grey and one brown. The red box, at 38 litres, is the smallest. Into it go paper, junk mail and textiles. The blue box, at 55 litres, is nearly 50 per cent bigger. Into it go plastic bottles, steel and aluminium cans, glass bottles and jars. Another important difference is that the boxes are not only emptied by the bin-men each week, but sorted by them too. The paper and textiles from the red boxes are separated into their own compartments on the collection vehicle (you could hardly call it a dustcart). The glass from the blue box is separated from the cans and plastic bottles, and sorted by colour – clear, green and brown. Back at the depot each material is tipped into its own designated bay, whence it is shipped by articulated lorry for recycling. The standard grey wheelie bin is for ordinary domestic refuse, which goes to landfill. The brown one is for organic garden waste and cardboard, which goes to a local composting site, run by Biffa, whose output is used as a soil improver in the landscaping of landfills. All this enables Daventry to claim a high (for the UK) recycling rate of 42 per cent, though critics point out that 26 per cent of the total is achieved through composting and not materials recycling. In 2005–6, when it has a statutory target of 36 per cent, it expects actually to achieve 45 per cent, with a long-term ambition of 50 per cent. Against this my own local authority, North Norfolk, could claim only 29 per cent and faced a supreme effort in getting up to the 36 per cent target for 2005–6 – an effort that would depend on its new wheelie bins and contract with NEWS.

Both authorities are convinced of the rightness of their methods. NEWS rejected kerb-sort specifically because of its high cost (Daventry spends £65 a year per household, against

North Norfolk's estimated £40), which it felt could not be justified by the undeniably cleaner, better-sorted material that would come into the plant. Unlike the Norfolk districts, Daventry finds itself in the top 25 per cent for performance. But unlike them, too, it is among the worst 25 per cent for cost and has attracted criticism from the Audit Commission. The council nevertheless remains steadfast. It 'does not intend to be deflected' by criticism, and predicts that other councils' costs will rise as they, too, begin to engage with their statutory targets.

In a way, such disagreements are encouraging. The argument is not about whether or not to recycle, but about how to achieve the best results. The overall 'best practice' models will emerge over time, and good forward planning (exemplified by both NEWS and Daventry) allows enough flexibility in the system to respond to new or changing markets. It is the market, however, that calls the tune. You cannot recycle material for which there is no practical purpose and which nobody wants to buy. That way you simply convert one kind of waste into another. Systems have to be founded on a bedrock of commercial confidence. To attract investment in plants such as NEWS, the markets have to be long-term, and stable enough to offer price guarantees. NEWS itself has contracts with its customers extending for up to seven years, and has been able to operate profitably from day one.

The downside is that rubbish without a market will never be anything but garbage. From this perspective, the zero waste economy looks about as achievable as the canonization of Pol Pot. Where there's muck, there's muck. 'There will always be landfill,' say the pragmatists, and you wouldn't bet on them being wrong. On one issue after another, the industry waits on the government for a lead. Companies won't invest until they know what the statutory requirements are going to be; can't risk a top-of-the-range specification if cheaper outfits with lower standards will be allowed to undercut them. If landfill is to be replaced by other methods, then standards will have to be set. With anaerobic digesters, for example, much hinges on what happens to the dry, inert material that comes out at the end. Will it be landfilled? Or could it be used, perhaps, for landspread to help farmers replace their wasted soil? And so on. Time ticks. The industry waits.

It is not just through domestic refuse collection that materials are sorted and sent for recycling. As we have seen, WEEE, ELV, tyres and batteries are finding dedicated routes of their own. The scrap metal business continues to work in its own mysterious way. Household waste recycling centres mop up the stuff the dustmen can't take. In industries of all kinds, one man's junk has long been another man's goldmine. Furniture increasingly is made from compressed bits of dust and twig for which previous generations would have had no use beyond the smoking of kippers. Fleecy jackets may have begun life as plastic bottles. Marmite famously recycles a residue from the brewing trade.

Old plastics, glass, paper, aluminium, wood, oil, textiles and steel all have well-established uses and there is an urgent need for more. In 2001, as part of Waste Strategy 2000, the government allocated £40m, spread over three years, to a specially created, cutting-edge recycling think-tank and promotion unit, the Waste Resources Action Programme, known (you guessed it) as Wrap, based at Banbury. Its remit would have done credit to a mythological king ordering miracles from a wizard. With one wave of its wand it had to create 'stable and efficient markets for recycled materials and products for the 100 million tonnes of waste accounted for by commercial, industrial and municipal waste'. With the next, it had to concentrate its powers of transformation on six specific 'material streams' – aggregates, glass, organics, paper, plastics and wood.

In 2003, with the wand still in the air but the audience yet to see anything to gasp at, the job-list took a more prosaic turn. The prime minister's strategy unit by now had nosed its way around the issues and become a Concerned Shopper. Finding new uses for old rubbish was all very well, it seemed to think, but we shouldn't appear to be legitimizing the creation of waste, we should be trying to minimize it. Back went the wand in the opposite direction. The next trick would be 'reducing waste at home, by working towards increasing home composting, reducing nappy waste, working with the big retailers to reduce supermarket waste and creating a waste minimisation research fund . . .' Wrap also had to provide a recycling advisory service for local authorities, with 'support for the composting industry to expand to absorb the extra material collected'. Simultaneously it was to 'engage' the public 'by raising awareness of the need to reduce waste and recycle more, particularly by helping councils get the most out of their collection schemes by promoting them effectively'. Your own state of awareness – heightened or otherwise – will give you some idea of how well this has worked. In some areas it seems not only to have failed to work but actually to have been counter-productive. A local official in Norfolk, for example, complains that landfill tax credits siphoned away from the industry and given to Wrap have yielded no visible benefit. NEWS itself lost about a million pounds' worth of funding which has 'gone out of Norfolk permanently'. With it, they could have mounted a county-wide education campaign that might have brought at least the non-comprehenders and over-enthusiastic recyclers on side. Without it they couldn't, so the well-meant but pointless flow of yoghurt pots continues. I ask if the county has seen any benefit from Wrap at all.

'No. None whatsoever.'

Jealous eyes are cast at the huge increase in funding that arrived in the wake of the prime minister's strategy unit – another £15.1m to add to the £40m previously granted for the first three years, then £52.3m for 2004–5 and £84.4m for 2005–6. It makes an inviting target. There is derision for the Wrap recycling logo, characterized as 'the flying sperm', and a widespread view that the whole thing has been just political window-dressing – a too-little, too-late bit of environmental tokenism. 'The government had to be seen to be doing some-thing. Wrap was the means of doing it.' This may be a bit hard. While you could hardly accuse the government of taking waste too seriously, Wrap itself may deserve better than to be damned by association. Biffa's Peter Jones, a man with no fear of plain speaking, remains cautious. While taking a 'jury's still out' line, he agrees that Wrap is 'undoubtedly more focused in a shorter timeframe' than previous approaches based around landfill operators. Even rural North Norfolk has had *some* direct benefit. Wrap's self-proclaimed 'flagship project' has been to convert the UK's largest newsprint mill, operated by the Finnish papermaking giant UPM-Kymmene at its Shotton works in North Wales, to 100 per cent recycled fibre. This has thrown up some impressive numbers. By providing a recycling route for an extra 321,000 tonnes of paper, it allows another 4m homes to recycle their newspapers and magazines and means that all the newsprint in the UK is made from recycled paper. Shotton now has the biggest single 'de-inking' line in the world, its appetite extending to 620,000 tonnes of used newspapers and magazines a year. (And yes, they've worked out the bus equivalent – it comes to 82,666 double-deckers.) This translates into approximately 80 per cent to 85 per cent of its weight in fresh newsprint. One of the local authorities that have taken advantage of Shotton's long-term price guarantees is Norfolk County Council, through NEWS, which, by its own account, is more than pleased with its contracts. And if research and development projects sponsored by Wrap succeed in their stated aim of developing new products for new markets, then NEWS's own avowed aim of profitably recycling a wider range of materials will have been done no harm.

Would the development of Shotton have happened anyway, regardless of Wrap? Probably not. NGOs can only push at doors; they don't have the muscle to shift entire markets, but even unlatched doors need someone to nudge them open. Wrap can oil the hinges, create new lines of communication, accelerate the timetable. The development of Shotton cost £128m, of which £17m came in grants from Wrap – partly from its own pocket and partly under the tax credit scheme from landfill operators and their environmental bodies (EB Nationwide, Biffaward, SITA Environmental Trust, Cleanaway, Viridor Credits and WREN). In December 2004, Wrap gave another £10.6m towards a £51.7m, biomass-fuelled sludge boiler, due to be operational by the end of November 2006, which will use the plant's waste material – mainly inks and spent fibres – to generate heat and power. Carbon dioxide emissions will be reduced by 30,000 tonnes a year, and the plant will become more than 90 per cent self-sufficient in heat and up to 25 per cent in power.

Wrap's other greatest hits so far have included truck body panels made from recycled lightweight plastic that are 60 per cent lighter than conventional panels – sufficient to increase each vehicle's payload and cut out one lorry-trip in every eight or ten – and a 'Recycled Garden' that made the rounds of 2004's summer flower shows. In September 2004 it had a little push on the public awareness front when the BBC makeover show *Changing Rooms* devoted a programme to recycled home-improvement materials. The quadruple Olympic gold medal-winning oarsman Matthew Pinsent climbed into a motorized can to launch what was heralded as 'the first [i.e., long overdue] national advertising campaign . . . in England to encourage people to recycle more household rubbish more often'. A series of television ads, featuring Eddie Izzard, showed how metal cans, glass and paper could be recycled into 'interesting new everyday items'. Well, it's a start. And, for those whose interest tipped into enthusiasm, there was a website, www.recyclenow.com, with genuinely useful information about where and what to recycle, and with a link to a sister site listing more than 3,000 recycled products already on the market. It's not the sexiest come-on you'll ever fall for, but there's no denying the force of its example. There are tyres, cat litter, carpets, traffic bollards, park benches, scissors, fencing, cycle stands, signposts, cushion covers, garden sheds, manhole covers, bar stools, car parts, pillows, litter bins, animal bedding, ballpoint pens, seed trays, hosepipes, coat pegs, pallets, clocks, tableware, clothing, soap dishes, paving slabs, lighting, loft insulation, water butts, printer cartridges, bin liners, just about anything you can think of made of paper. And coffins.

Some of it is small-scale; some of it reflects major commitment from industrial heavyweights. But still it's no one's Klondike: for all the improving technology, all the cool talk and ritzy logos, the great Recycling Rush has yet to catch fire. No one, yet, has managed to make it look sexy. Worthy, yes; sensible, certainly. But seductive? The very words, 'Made from recycled paper', look more like an ideological statement than a mark of quality. You could read it almost as an apology – this stuff may not be glossy but, hey!, we're doing our bit for the grandchildren. The real breakthrough will come when the words are no longer needed; when recycling is as unremarkable a fact of industrial life as the chairman's reserved parking place. It will require the same breed of hero bringing the same kind of enterprise and imagination that drove the industrial and electronic revolutions of the nineteenth and twentieth centuries. Like them, too, it will need investment; and bold ambition driven by good old-fashioned commercial competitiveness, not targets from government offices. In October 2004, Wrap reported that 81 per cent of the UK's recycling businesses believed Britain would be unable to meet its EU recycling targets unless there was more external investment. The same survey revealed that 40 per cent of the companies were being held back by lack of capital. Wrap itself reckoned that the UK would need an extra £6 billion investment in reprocessing capacity alone, just to meet its targets.

In the meantime, its six 'material streams' all flow in the direction of hope. With glass, the major problem remains, as it always has been, colour. Half the glass from bottle banks is green – most of it in the form of imported wine bottles for which there is no manufacturer, and hence no market, in the UK, where food and drinks manufacturers mostly prefer clear. Some of the green can be crushed and used as a substitute for limestone chippings in road-building aggregate, or in place of sand to filter swimming pools and water purification plants, or as an ingredient in concrete and ceramics. A stylish kitchen work surface called TTURA, made from recycled glass, drew admiring attention at the 2004 Ideal Home Show. It all amounts to a step, rather than a surge, in the right direction. Wrap itself has done a bit of arithmetic based on the Packaging Waste Regulations. With the glass target for 2008 set at 60 per cent, the recycling rate would have to double in four years from 2004 to an approximate annual total of 1.38m tonnes. While the glass industry can use quite a lot of this as 'cullet' – waste for melting down – it still leaves a potential surplus of 400,000 tonnes of green glass looking for a home. Wrap's answer has been a £48,000 investment in research carried out by the Sheffield-based technical consultants Glass Technology Services with the engineering materials department at the University of Sheffield. The challenge: to find a way to neutralize colour in the furnace and turn green glass into clear. Can it be done? In June 2004, after the researchers had tried six different techniques, Wrap issued a statement that looked more like hope than conviction, but at least it was a hope to which we could all raise a glass: 'Following publication of the report, discussions are being held with the glass container manufacturing industry to examine the findings and establish how this research can be taken forward.' In October 2004 it pressed on with a new project to increase the amount of green used in the manufacture of clear and amber glass, and to test public acceptance of clear glass containers with a 'greenish tint'. It is an issue of major economic importance for local authorities, which have to decide whether or not to colour-separate the glass they collect (a process requiring significant investment) before they sell it on for reprocessing.

The Concerned Shopper, of course, has an awkward question or two. If green glass is such an awful problem, then why do we keep on importing it? Why can we not persuade wine producers to send the stuff in clear glass? Answer: there is about as much chance of persuading the vignerons of Bordeaux and Burgundy to change their bottle colour (which they insist is essential for the protection of the wine) as there is of turning them away from Cabernet Sauvignon or Pinot Noir and on to rhubarb and parsnip. That being the case, why do we not bow to the inevitable and use green glass ourselves? Answer: we're trying.

Not all recyclable glass comes from bottles and jars. Wrap also has ambitions for window glass – it wants new infra-structure to collect and process another 50,000 tonnes a year – and for cathode ray tubes and lighting glass, for which it envisages 'closed loop' systems creating new from old. Other ideas range from the usefully dull to the almost wacky: glass as a filtration medium for processing industrial waste; glass grit for shot-blasting; ground glass mixed with clay in bricks (a 10 per cent glass content means a 20 per cent energy saving, it says); glass 'sand' in golf bunkers.

If the future of glass in many cases is to be replaced by plastic, then perhaps it should be no surprise that plastic, too, throws up problems of colour. Coloured bottle-plastic, known as Jazz PET, in all but colour is identical to 'natural' PET but has only 30 per cent of its market value. London Metropolitan University is looking for ways to use Jazz PET in the manufacture of guttering, drainpipes and coathangers – all potential long-term markets unlikely to be influenced by fashion.

The sorting problem is being addressed too. Wrap has given more than £1m – approximately a third of the total project cost – to JFC Delleve at St Helens to help it develop the most technically sophisticated plastic-bottle sorting and recycling centre in Britain. When it gets up to full speed in July 2005, it will be able to recycle 20,000 tonnes of plastic a year. This is the equivalent of 500m individual containers, approximately half of which will be PET (water and fizzy drinks bottles) and half HDPE (milk and juice). Technical sophistication comes in the form of a hot wash system that will get rid of all residues and contaminants, including glue (which is immune to the current cold wash), and an automatic separation process, similar to the one at NEWS, that will use infra-red technology to sort the PET by polymer type and colour. The colour-sort is necessary because the principal use of recycled PET is as 'feedstock' for polyester fibre, and only clear or blue PET can be made into the white fibre that most manufacturers want. Green and other colours make a black material that can be used only in places where appearance doesn't matter – in duvet linings, for example, or under-bonnet insulation in cars. The most important current uses for recycled PET are textile 'fleeces', duvets, pillows, coat linings, carpet fibres, car door linings and boot trays.

'Most important', however, does not mean 'most bottles'. In terms of the overall recycling rate, all this makes hardly a dent in the mountain. The UK produces more than 450,000 tonnes of plastic bottles a year, of which a meagre 5 per cent is collected for recycling – approximately half of it in the UK, with the rest being exported for processing abroad, mainly in China and India. The remaining 95 per cent goes uselessly to landfill. Unsurprisingly, Wrap is agitating to increase the volume of recycled bottle-plastic used by the packaging industry (an ambition it shares with the processors), and to identify and develop more uses for recycled plastics generally.

In the wider market, one promising development, yet to catch on, is the substitution of recycled plastic for wood. It can be made to look like the real thing, and, like the real thing, can be supplied as planks, poles or sheets. Unlike the real thing, it is durable, weatherproof, tough and maintenance-free. When used outdoors – for seating, footbridges, jetties and decking, for example – it has four times the life expectancy of natural wood. Wood itself is a problem for recyclers, principally the particleboard industry, because of its con-tamination by varnish and preservatives. Ironically, the more care you take to protect natural wood, the less likely it is to enjoy a useful afterlife. No old plank or panel is ever going to come with a written provenance, and this is no place for guesswork. Wrap therefore is looking for a technique to identify preservatives, and investigating the possible use of infra-red scanning to measure the varnish.

Wrap's overall contribution to the recycling economy is difficult to evaluate. It has sucked up funds from the landfill tax credit scheme, and has moved into the territory of other semi-official bodies and environmental charities – Global Action Plan, Waste Watch, the Environment Council and others – so it is impossible to deduce exactly how much it has done that would not have been done anyway. Impossible calculations by their nature yield unreliable answers. In the case of Wrap, conclusions are heavily influenced by point of view. Those who feel they have lost out have a different perspective from those who feel they have gained. Others, like Peter Jones, wait in the jury room. It is, however, permissible to hope, for in many ways Wrap is the only show in town. Waste handling is a business, profitable for some, but it is a risky one, and rare is the investor who feels obliged to take a punt on problems that are not of his own making. Wrap's handouts cut the risk and tip the balance. All council tax payers potentially stand to benefit from its priming of the pump and long-term expansion of the market. This alone, if properly understood, should freeze the tongues of those, stake-holders all, who complain about having to sort their garbage. Perhaps more importantly even than this is Wrap's crucial function as the hinge between research and application, matching ideas to money and environmental need to commercial opportunity. I yield to no one in my suspicion of government-inspired initiatives with messianic mission statements. I yield to no one in my hope that, this one time, they'll get it right.

Getting it right means more than just meeting recycling targets. The problem with targets is that they cut out peripheral vision, obscure their own purpose and become the ends that justify the means. A prime example was the obsession with hospital waiting lists. Short waits good, long waits bad – it looks, and in many ways is, a reasonable assumption: shorter waits mean faster treatment, less pain, better chance of cure. But hospital waiting lists are not like supermarket queues. Not everyone shuffles forward at the same speed. You move forward or back according to the severity of your case (ironically inviting statisticians to conclude that short waits are bad for your health). Short queues can mean all kinds of things. GPs may not be bothering to refer patients who have no chance of treatment; or they may be doing an exemplary job in keeping them in good health. It may be that they have a young and fit population. The hospitals themselves may be super-efficient, or they may be under such intolerable pressure that they are simply pushing patients through the system faster than is good for them.

Wherever targets compete, you get trade-offs. If a hospital is doing badly on waiting lists, one solution might be to shorten the length of stay in hospital. It would be good for the waiting list but might not be quite so good for the patient. You can see it in the statistics. In a piece for the *Sunday Times Magazine* in May 2000, with the help of the King's Fund I looked at some of the trends behind the figures. One example was an NHS Trust in the North-East which had topped the national chart for speed of admissions, and whose GPs were the champion prescribers of generic rather than (much more expensive) branded drugs. Viewed on these criteria alone, it looked a model of good practice. Yet what happened when you looked at clinical outcomes? It dropped to 67th place for needless hospital admissions, 80th for avoidable deaths, 94th for fatal heart disease and stroke, 96th for overall mortality and 97th for cancer deaths. What did it prove? That political emphasis on cutting waiting lists was just that – politics. The real cause of the area's ills was to be found in the Under Privileged Areas (UPA) index, which measures social deprivation, where it ranked 84th. Meeting waiting-list targets for the urban poor is about as soothing to their symptoms as wrapping their prescriptions in prettier paper.

What has any of this got to do with recycling? Nothing, save to offer a specific example (there are many more, in crime prevention as well as health) of the way targets can lay a rose-coloured smokescreen. In the UK there is an imbalance between recycling targets, the infrastructure needed to meet them, and available markets for recycled materials. This is why Wrap found that most recycling businesses believed the targets could not be met. It is also the reason why the UK has kept up a steady flow of end-of-life exports – fridges, WEEE, batteries – to its better-equipped European partners.

The export trade, however, is a wide and stretchy carpet beneath which all manner of horrors can be swept. The classic way for a rich industrial nation to solve a problem is to export it to a poorer country on the other side of the globe. In a kind of grossed-up parody of the Victorian dust-yard, the world's poor sift through the leavings of the wealthy in search of marketable salvage. It is, or can be, a double benefit. Where the trade is legitimate (plastic bottles, for example), the exporting country can claim success towards its recycling targets while the host country, which does the actual recycling, profits from the sale of the recycled material. Where it is illegitimate (toxic waste), the exporters prey on the desperation of people whose poverty drives them to risk their health, their lives and their environment. It is a sophisticated modern adaptation of the caveman's proto-Nimbyism: don't do on your own doorstep what you could do on someone else's. You hit your targets either by recycling legally in another country, or by exporting the stuff illegally and 'vanishing' it – unless of course you are American, in which case the question of illegality seldom arises. From the US you can export more or less anything you like to whomsoever is willing to receive it. Freedom demands no less.

As we have seen, half the plastic bottles collected for recycling in the UK are exported, mainly to China, which pays a good price for them. They are part of a gathering stream: unconfirmed latest estimates (I am writing in early December 2004) suggest that Britain sends to China an annual total of 200,000 tonnes of waste plastic, of which between 10,000 and 15,000 tonnes are in the form of bottles, and 500,000 tonnes of paper and cardboard. It is hard to make any kind of moral or ethical case against this. The material is pretty much hazard-free. There is no coercion. One party has something to sell which the other wants to buy, and why shouldn't they? Where there are swings, however, the roundabouts are never far to seek. China has cheap labour and can offer much higher prices than UK recycling companies can afford – even mixed plastic bottles, for which there is little or no market in the UK, can fetch up to £50 a tonne. The result is a high casualty rate in British businesses, a powerful disincentive to significant new investment, and a major obstacle in the way of Wrap's wholly laudable aim of accelerating the local recycling rate. Result? Short-term targets met, long-term ambitions thwarted, and a waste economy that sends vast consignments sailing wastefully around the globe.

Albeit with regret, you can make a kind of hard-headed sense of this (trade will always flow in the direction of profit), but those with anything but a moral vacuum at their heart will find no such justification for the illegal export of toxic waste. Human misery is, or ought to be, untradeable. The European Union has banned the export of hazardous material beyond its own borders (hence the fridge mountain, and desperate anxieties over the lack of UK processing capacity). The United States has no such scruple. The consequences of its rich man's incontinence were set out, in agonizing detail, in a report published in February 2004 by an international coalition of environmental organizations led by the Seattle-based Basel Action Network (BAN). It found that vast consignments of old computers and other hazardous electronic waste, mainly but not exclusively from North America, were being shipped to China, Pakistan and India, where they were being 'recycled' by workers in conditions that would have appalled a Victorian gangmaster. At Guiyu, on the Lianjiang river in Guangdong Province, it came upon a colony of 100,000 migrant workers – men, women and children – living what Jim Puckett, BAN's project co-ordinator, called a 'cyber-age nightmare'.

'They call this recycling,' he said, 'but it's really dumping by another name.' There are no health or environmental controls. Plastics and wires are burned in the open; soldered circuit boards are melted and burned; lead-contaminated cathode ray tubes are dumped. Observers reported seeing women and girls heating lead solder in woks over open fires, then using the molten solder to loosen memory chips from computer circuit boards. Afterwards the used lead, universally recognized as a neurological toxin, is tipped like kitchen-slops on to the ground. The residue of material – metal, plastic, chemical – that has no value is dumped in fields and irrigation canals, with the result that local well water is polluted and fresh water has to be trucked in from elsewhere. To no one's great surprise, the US is the only industrial nation in the world not to have ratified the Basel Convention, a UN treaty that seeks to impose a worldwide ban on the export of hazardous waste from rich countries to poor (it is from this treaty that the Basel Action Network takes its name). It is not the case that the US has no laws governing exports. It does, but specifically exempts electronic waste on the ground that it is being sent for 'recycling'. 'To our horror,' said Puckett, 'we . . . discovered that rather than banning it, the United States government is actually encouraging this ugly trade in order to avoid finding real solutions to the massive tide of obsolete computer waste generated in the US daily.'

Bush's America, where 'yee-haw' seems to serve just as well for environmental policy as it has done for foreign, is an easy target for European finger-pointers. The biggest, muckiest and most powerful brute in the international playground, it has no need of subterfuge and can openly do what it likes. One US-based broker boasted that he exported more than 10,000 computers and monitors every month. 'We export to every major country in the world,' he said. 'We regularly ship to Pakistan, India, Sri Lanka, Egypt, Iran, Bangladesh, Azerbaijan, Vietnam, Jordan, Iraq, Saudi Arabia, Kuwait, Syria, Germany, Spain, France, Bulgaria, Italy, Belgium, Russia, Portugal, Romania, Ukraine, Poland, Canada, Hong Kong, Bolivia, Venezuela, Colombia, Peru, Chile, Nigeria and Ghana.'

Our own transgressions are more clandestine, on the wrong side of the law and less easy to quantify. At least some of the toxic e-waste polluting land and water, and endangering life in societies less privileged than our own, was illegally exported from the UK. The difficulty lies in knowing how much. In December 2004 a report compiled for the Environment Agency by the Industry Council for Electronic Equipment Recycling (ICER), the representative body of the WEEE recycling industry, suggested that 23,000 tonnes of electronic equipment, including at least 10,000 tonnes of used PC monitors, had been illicitly shipped out during 2003, mainly to

China, the Indian subcontinent and West Africa. The problem is that much of the junk, which includes millions of mobile phones, masquerades as legal trade in workable second-hand equipment and either passes undetected or is given the benefit of the doubt by port inspectors who can't tell the difference between reusable equipment and scrap. Sometimes the deceptions involve long and complicated paper trails designed to confuse, often with reroutings via other ports (Rotterdam, Gibraltar and the Middle East, especially Dubai, are particularly popular). Sometimes they are almost mockingly crude. A container of 'plastic' awaiting shipment from Felixstowe to Lahore on inspection turned out to be a rubble of computer monitors and other electronic waste on its way to Pakistan from South Wales. As legal disposal in the UK becomes both more difficult and more expensive – it can cost £5 to recycle a single computer monitor here – so yet again the cowboys are saddling up and the pirates putting on their eyepatches.

At least now they will have to look over their shoulders. Following the ICER report in late 2004, the Environment Agency announced an international crackdown, involving enforcement agencies in twelve European countries with jurisdiction over twenty-five major ports. In the UK, the agency itself would set up an enforcement team to greatly increase the number and frequency of inspections at ports in England and Wales, and to improve the likelihood of offenders being caught and prosecuted. An earlier mini-crackdown, involving six European countries between spring 2003 and summer 2004, found that a shocking 20 per cent of inspected waste shipments were illegal.

There is money in such trade, and an intoxicating whiff of

danger, but no romance. Environmental crime is not victimless; the acts of its perpetrators are little different in effect from firing bullets into a crowd, or putting poison in the water. If we want to set targets, then we should be ready to meet them, legally and safely, from our own resources. It is infrastructure we need, not just more environmental policemen. For as long as we fail, we sustain the human equivalent of the NEWS salmon leap – except that, in a moral vacuum, the polarities are reversed. Decency is canned, and the rubbish jumps for joy.

CHAPTER NINE
A New Kind of Junk

S TRANGE THINGS HAPPEN. ONE FREEZING WINTER NIGHT I find myself at dinner in a Norfolk farmhouse. The guests are an eclectic mix, from a variety of educational and professional backgrounds, mostly strangers to each other and with nothing in common save acquaintance with our host. Canapés are chased down by pheasant, treacle tart and syllabub. Pinot Noir follows Sauvignon Blanc; Old World follows New. The talk grows louder, bolder, less discreet. At the end of it all, our departures are delayed when someone produces a bottle of chilled Russian vodka, flavoured with blackcurrant. So far, so very NFN (Normal for Norfolk). But then someone else in a distant room puts on a CD and the house at midnight floods with music, like a cathedral. The laughter is stilled. This is not a religious gathering (no churchgoers that I am aware of), and we don't know until we're told that what we hear are hymns and psalms from the Night Office of the Russian Orthodox Church. The settings are by Vasily Titov, Rachmaninov and other Russian composers whose names are less familiar; the voices rich and dark, like carved oak translated for the ear. With our arms halfway into our coats, we are silenced, transfixed; tiredness and baby-sitters suddenly unimportant.

No one is surprised. We may vary in our standing with God, but we all know what music can do. A week later we are in Norwich for the Glyndebourne Touring Opera's beautifully sung *La Bohème*. The Theatre Royal has a big auditorium (more than 1,300 seats), but the place has been packed all week. If you hadn't booked early, then you'd have had to wait until 2005. It is a kind of affirmation – confirmation, as if any were needed, that good things hold their value. We want uplift in our lives, and we'll drive through the fog to get it. It doesn't have to be opera. Back in the summer it was Jools Holland's rhythm and blues orchestra. It can be Shakespeare at Stratford; a raw rock band; good jazz; good film; good theatre. Tastes vary, but not the appreciation of quality. I presume this to be true even of rap.

But what of television? My dinner companions used their sets, if they still had them, for video or DVD, and hardly at all for the 'rubbish' listed in the *Radio Times*. As artists never tire of reminding us, no worthwhile advance in cultural life ever comes without opposition from conservative diehards or moral absolutists who can see no point of view but their own. Picasso was ridiculed, but is now revered as a genius. Tracey Emin is ridiculed, ergo she is a genius too. This may be a whole premiss short of a syllogism, but we can take the point. 'New' and 'different' are not synonyms for 'worse'. Rubbish is in the eye of the beholder.

And yet value judgements are not only irresistible but so consistently in agreement that they transmute into objective truths. Television is the great polarizer. As more and more people are turned off by what they are being offered, so more and more are drawn to watch. Here is Alan Hamilton, writing in *The Times* in August 2004:

> Like drugs, there is money in lowest-common-denominator vacuity. Endemol, the independent producer which makes *Big Brother* alongside its highbrow output of *Changing Rooms*, *Ground Force* and *Ready, Steady, Cook*, last year posted pre-tax profits of £15.8 million on a turnover of £89 million. No one ever lost a fortune by underestimating the public appetite for tackiness and vulgarity.
>
> The satanic genius credited with introducing the idea of *Big Brother* to Britain is Peter Bazalgette, the Cambridge-educated son of a stockbroker who heads Endemol. He is the great-grandson of the illustrious Victorian engineer Sir Joseph Bazalgette, who built London's sewers and rid the capital of a great stink. There are those who think that the present generation has reversed the flow.

Later the same month, the broadcaster John Humphrys delivered an acerbic MacTaggart Lecture at the Edinburgh Television Festival. It had been five years since he last owned a television, so he'd had a lot to catch up on. He, too, had been shocked by the calculated cynicism of shows like *Big Brother*; by the habit of accepting audience figures as the single meaningful measure of a programme's worth; by the desperate upping of the exploitative ante, necessary to hang on to audiences so coarsened and desensitized that they needed an ever-stronger diet of shocks to nail their attention. The defence of the channel chiefs is that television exists in a kind of cultural lotus-land where, irrespective of experience or education, we can all blissfully graze at the same perpetually fruiting tree. Television is neither high-brow nor low-brow, but 'no-brow'. By this reckoning, *Big Brother* is like the taste of chocolate, as delectable to a Nobel prizewinner as it might be to a dustman. Humphrys rejected this. It was no-brow, he said, 'because it is no content, no nourishment, no good'. With the unwavering emphasis on 'entertainment', commissioning editors had become risk-averse and originality was not required.

Thus it might be deduced that the trouble with television may not be too much rubbish but not enough. Triumphs are conceived by courage out of originality. But originality always carries risk, which means we must accept some heroic failures too. To create and talk rubbish is an essential human freedom, our defence against 'official taste', and we need more of it, not less. The real problem with much (not all) of the broadcast output is not that it damages public morals but rather that it is an unadventurous, production-line commodity that is the cultural equivalent of sliced bread. Even in its poor taste and nihilistic viciousness, it only follows a recipe. Homogeneous, superficial, briefly entertaining but instantly forgettable, a colossal, jobsworth-funding energy-waster, it is the very paradigm of twenty-first-century junk. At least it doesn't stick to your shoe.

In December 2004, with the renewal date for its royal charter only two years distant, the BBC launched another of its periodic self-improvement exercises. Its new director-general, Mark Thompson, announced 2,900 job cuts and a 15 per cent across-the-board reduction in departmental budgets. Somehow, by hacking into dead wood, trimming the fat and dropping the passengers, this would all translate into better programming – more big-budget dramas, big 'factual event shows', beefed-up news and current affairs, and wholesome family entertainments like *Strictly Come Dancing*. There would be fewer repeats, less dross from across the Atlantic, and a principled turning aside from 'reality' and ticky-tacky DIY makeover shows. A minimum 25 per cent of programmes other than news would be bought in, or 'outsourced', from independent producers, and various of the currently London-based departments – sport, children's programmes, new media, Radio Five Live, 1,800 jobs in all – would shift to Manchester. The news was better received by estate agents in the North-West than in Broadcasting House, where staff morale – still raw from the Hutton fiasco and the enforced departure of the popular Greg Dyke – had reached an historical nadir. Do demoralized staff make better programmes? Stay tuned.

The blurring of language is another price we are asked to pay for the convenience of mass communication. In the age of the text message, there are people with impressive academic titles who are willing solemnly to argue that spelling and grammar are unnecessary barriers to the information superhighway, being manned by the same white, university-educated, patronizing middle-aged Home Counties buffers who want to see stuff like opera and Shakespeare on TV. Wel bd lk 2 m. Our history is in our language, just as it is in our old buildings and the ancient pattern of streets. They have an aesthetic value as well as a function, and it is no coincidence that the institutions of government are as clumsy with words as they are with the placement of their housing quotas. The following examples are all 'Golden Bull' prizewinners chosen by the Plain English Campaign:

RUBBISH!

From the Department for the Environment, Transport and the Regions, 1999:

In the application by virtue of this paragraph of subparagraphs (4) and (6) to (10) of paragraph 3 to an application or proposed variation:

(a) the notice served under sub-paragraph (6) of this paragraph shall be treated as the notification required by sub-paragraph (4) (1) of paragraph 3;

the reference in sub-paragraph (6) of paragraph 3 to the day on which the notification under sub-paragraph (4) a) of paragraph 3 is made shall be treated as reference to the day on which the notice served under sub-paragraph (2) of this paragraph is given.

From the Department of Trade and Industry, 1999:

a person carrying on an employment business shall not request or directly or indirectly receive any fee from a second person for providing services (whether by the provision of inform-ation or otherwise) for the purposes of finding or seeking to find a third person, with a view to the second person becoming employed by the first person and acting for and under the control of the third person.

From the Department for Education and Skills, 2001:

Paragraph (1) is without prejudice to any action which the Corporation may take in relation to a clerk who is also a member of the staff by way of suspension from or termination of the appointment as clerk under the terms of any separate appointment as clerk.

From the Department of Health, 2004:
'Container', in relation to an investigational medicinal product, means the bottle, jar, box, packet or other receptacle which contains or is to contain it, not being a capsule, cachet or other article in which the product is or is to be administered, and where any such receptacle is or is to be contained in another such receptacle, includes the former but does not include the latter receptacle.

The government is not the only, or even the worst, offender. Here is a recent attempt by British Airways to explain its terms and conditions:

Charges for Changes and Cancellations
Note – Cancellations – Before departure fare is refundable. If combining a non-refundable fare with a refundable fare only the y/c/j-class half return amount can be refunded. After departure fare is refundable. If combining a non-refundable fare with a refundable fare refund the difference/if any/ between the fare paid and the applicable normal BA one-way fare.
 Changes/upgrades – permitted anytime.

This is from the 'Genius project' at the University of Reading:

The project is structured around multifaceted incremental work plan combining novel content design based on new pedagogical paradigms blended with the e-learning environments to facilitate hybrid mode of delivery. This is combined with series of educational experiments on the target learner groups with possibilities to adjust the approach and disseminate the interim and final results.

Our pedagogical approach is based on the educational model which assumes that the learning process is an interactive process of seeking understanding, consisting of three fundamental components: Conceptualization, Construction and Dialogue. The relevant modules of the New Curricula are mapped onto these three components and a hybrid way of delivery is investigated through different scenarios.

Four government departments, a flagship company, a university – unfairly singled out, perhaps. They could argue that there are many others just as bad, and they would be right, but we would find no consolation in it. We would prefer it if they were exceptional. Time and again, when we enter the labyrinths of European and domestic policy, we are trapped by dense thickets of dubious meaning. In many cases not even the civil servants can understand each other. Not one I spoke to expressed the slightest surprise that I could make no sense of raw European directives. They couldn't either (though some were prepared, as one of them put it, to 'have a go'). Here is the entire explanation for the fridge fiasco – London simply did not know what Brussels was talking about. Here, too, is a large part of the reason for the waste and retail industries' failure to be ready for the new European directives – fridges, tyres, hazardous waste, WEEE, batteries . . . They ask straight questions of the policy-makers but get no straight answers (often they get no answers at all). Yes, language does matter. It is a fundamental tool of democracy, a window into the minds of our political leaders, which needs to be kept clean, transparent and free from fog. Sending spelling and grammar to the recycle bin is not 'empowering', as linguistic progressives like to claim – unless, that is, they mean only the empowerment of obfuscation and spin.

In waste-reduction as in all things, we are entitled to expect the government to lead by example. But we must be prepared for disappointment. It's all very well for it to tell us to sort our garbage and buy recycled paper, but how clean is its own shopping list? Biffa's Peter Jones treats the question almost as a joke. 'The government spends around 45 per cent of GDP, and 25 per cent of jobs are in the public sector,' he says. 'Yet in things like energy, building specifications, lighting systems and so on, there seems to be no strategic framework. Promoting such initiatives in schools, hospitals and offices would be a start. If consumers are expected to make an effort to buy less environmentally damaging goods, then so should the government. The message is undermined if the government fails to take its own advice.'

There was a sad but salutary example when Michael Meacher went to Manchester to open a recycling unit for fluorescent light tubes. Like the good and committed minister he was, he promised a million recycled tubes from government offices. Like the muddled and complacent bureaucracies they seemed determined to make themselves appear, the offices failed to deliver. The whole issue of 'government waste' is part of the political knockabout that will precede any general election. Everyone suddenly wants to cut 'back-office' staff and 'put more policemen on our streets'. One is deafened by the grinding of axes. In his July spending review in 2004, Gordon Brown said he would cut 104,000 civil service jobs across the UK. At least, most people thought that was what he meant – what he actually announced was 'a gross reduction in civil service posts of 84,150', leaving it to others to translate 'posts' into 'jobs' and do the sum. This provoked outrage from the civil service unions, supported by the Citizens' Advice Bureaux and TUC among others, who argued that the cuts would damage public services and led their members out on strike. Outrage, too, from the shadow chancellor Oliver Letwin, who complained of Labour's 'abject failure to slim down its "fat government"'. In October he published an analysis of the *Guardian*'s jobs pages, showing that 2,900 public sector jobs had been advertised there since Gordon Brown's announcement only three months earlier. 'If this level of recruitment were to continue unchecked,' said the Conservatives, 'Labour's fat government would grow by over 40,000 jobs at a cost of £1,575 million by 2008.'

A few days later, Mr Letwin complained that in the twelve months from April 2003, the Civil Service had expanded by a number equivalent to the population of Ilfracombe (12,280) to a total of 523,580 (Sheffield and Ilfracombe added together), though disappointingly he did not calculate their weight in double-decker buses. The recently formed Taxpayers' Alliance was also cross. It 'slammed' the government's 'high-spending agenda' and complained that the striking members of the Public and Commercial Services Union were 'overpaid and under worked', and that they took too much sick leave (an average of ten days, or two whole working weeks a year, against only seven days in the private sector).

The one thing this proves is that government waste, like the beauty of a butterfly, is not objectively quantifiable. It's not just a question of failing to chase the best deal on light bulbs, memo pads or police stations. One man's waste, as we have so often seen, is another man's bounty. To one observer, spending £30,000 on Pugin wallpaper for the lord chancellor's lodgings is a scandal. To another, it is essential investment in the nation's heritage. It all depends on where you stand. Stir in all the variables within the education, arts, science, transport, defence, health, work and pensions, environment and other departments of state, and you have more value judgements than you have clauses in a European Council directive. If you agree with a policy, it is money well spent; if you disagree, it's waste. Public funding for opera? Bah! Spatial development manager for a Lancashire borough council? Inverted pyramid of piffle!

In its *Bumper Book of Government Waste and Useless Spending*, which it published on its website in February 2004, the Taxpayers' Alliance was talking big numbers:

At least £50bn of taxpayers' money was wasted or spent on useless projects by the government in 2003. We have identified and listed more than 500 examples of spending that can be cut without closing a single hospital, firing a single teacher or cutting pensions. This is enough to abolish both council tax and corporation tax completely, or to chop 10p off each rate of income tax.

It lined up some familiar targets, headed by the 'useless' DTI, the abolition of which, it reckoned, would save £6.5bn a year. Other examples included Whitehall bureaucrats (£4.1bn wasted in 2003), benefit fraud (£3.2bn), Network Rail (£2.5bn), the Common Agricultural Policy (£1.8bn), defence write-offs (£1.7bn), local education authorities (£1.1bn),

GCHQ (£550m), Eurofighter (£360m), NHS administrators (£300m), government IT projects (£250m), mistakes by the Inland Revenue (£150m), the Bloody Sunday inquiry (£130m), the Common Fisheries Policy (£110m) and much, much more, all adding up to the headline total of £50bn. Some of this you may instinctively applaud; some you may be persuaded by; some you may reject. This is politics. A slightly more modest but still alarming annual figure is put forward in a working paper by Tim Ambler, a senior fellow of the London Business School, who makes a businessman's case – no less controversial for being well argued – for privatizing a (very) large part of the national estate.

'Each year,' he says, 'the UK government wastes about 8 per cent of its budget, or £30.7 billion in 2005, largely through bureaucracy and inefficiency. Determined leadership could downsize government and redirect most of this waste towards improved delivery of public services or, failing that, reduced taxes.' He recognizes the smokescreen laid down by politicians on the stump: 'In 2000, the Conservative Party claimed they could achieve annual savings of at least £8bn. Their failure to substantiate the arithmetic raised the question of how much really could be saved.' Ambler himself leaves no bean uncounted. In micro-economic detail, almost room by room, he works his way through each department of state, its satellites, quangos and funded NGOs. Functions are assessed and matched to headcounts; institutions are merged, abolished or privatized; headcounts cut (he notes in passing that the Cabinet Office has more than doubled its personnel since 1999 and is now twice the size of the US Office of the President). By trimming, and (especially) privatizing great lumps of the Ministry of Defence that most of us have never heard of (the Armed Forces Personnel Administration Agency, the Army Base Repair Agency, the Defence Aviation Repair Agency . . .), he lops 38,809 off the MoD payroll. Education and Employment loses 2,563, Foreign Office 1,926, Health 2,443, Home Office 3,711, and so on. Some satellite or part-funded institutions are nibbled at: the British Film Institute, for example, could lose 47, English Heritage 139, Historic Royal Palaces 51, Sir John Soane's Museum 2, Tate Gallery 58, English Nature 63, Coal Authority 99, Design Council 5, Particle Physics and Astronomy Research Council 47, Royal Marines Museum 2, Royal Botanic Gardens, Kew 56 . . . Others simply get the chop: the Apple and Pear Research Council, the British Potato Council, the Milk Development Council, the Commission for New Towns, even the Environment Agency, whose responsibilities he argues could be 'transferred to existing operations within local government'.

Well, hang on there. Whoa. Tim Ambler is not a man with whom I'd pick an argument over company balance sheets or any aspect of running a business – aside from his academic credentials he is a former joint managing director of International Distillers and Vintners. But environmental protection is not a business: it has no product you could put a barcode on; no service to sell. The business model is not always the right one. John Prescott's housing expansion plans and market renewal 'Pathfinders' show exactly what can go wrong when you hand social or environmental policy to economists. The Environment Agency is not as effective as we would like it to be – and certainly not as well funded. Its task as the nation's environmental enforcer grows with every new proposal, discussion paper, European waste or water directive and Act of Parliament, and its resources do not grow in parallel. The laws it enforces are nationwide; the industries it polices are nationwide; rivers and flood zones do not respect local authority boundaries; nor do environmental criminals. Against these, there has to be international co-operation as well as national co-ordination. Waste disposal, particularly of hazardous material, is a national and not a local issue, and it takes a powerful, well-funded professional agency to regulate it. Inland and coastal waters need to be under co-ordinated national control; so do flood defences; so do movements (especially cross-boundary movements) of hazardous waste. Only a specialist agency can have the range of expertise necessary to combine all these functions economically and effectively across wide geographical areas. It is far beyond the resources of local authorities to duplicate all this at local level – and they, too, as waste operators, need an agency to police them (Newcastle and Byker provide the example).

When government departments *do* attack waste, they tend to look at it through an ideological lens – clarifying or distorting, depending on angle of view. We search in vain for equilibrium, for a point of balance between the extremes of intervention and neglect. Once we were the world champions of transport and service infrastructure. Now we concentrate our housebuilding in the South-East of England – one of the driest parts of Europe – without knowing where the water is going to come from or the sewage go to, while large parts of the engineered and built environment are left to look after themselves, as if they were parts of the landscape. The reasons, again, are political. Replacing water mains costs money. Not replacing them causes bursts and leaks. But higher prices make bigger and worse headlines than burst water mains, so burst water mains is what we get. The worst example, endlessly chronicled and despaired over, is the railways.

As every boy used to know, the world speed record for steam locomotives, 126mph, was set on the east coast main line near Grantham on Sunday 3 July 1938. The engine was *Mallard*, a 103-ton streamliner designed by Sir Nigel Gresley, hauling a 240-ton, seven-coach passenger train of the London North Eastern Railway. The driver was Joe Duddington; his fireman Tommy Bray. Sixty-seven years later, their mark still stands. Scroll forward to Tuesday 16 August 2003. After a catalogue of disasters – the Paddington, Hatfield, Selby and Potter's Bar crashes, the public execution of Railtrack – it was a much-needed good news day for the railway. At Waterloo station, Tony Blair was preparing to open the UK's first stretch of high-speed line, a forty-six-mile section of the Channel Tunnel link between Folkestone and North Kent, that would allow Eurostar trains on British soil, if only for a few minutes, to nudge up to the continental standard of 186mph. On BBC radio that morning the transport secretary, Alistair Darling, had come to the microphone to lead the nation's cheers.

Shortly afterwards came news that an intercity express crawling out of King's Cross had run across an overnight track repair by the maintenance contractor Jarvis. The consequent derailment blocked six platforms and condemned the east coast main line to another two days of paralysis. It was not the only symptom of decline. Also falling silent, or at least a little quieter, were the skies over west London. Despite £17m worth of modifications designed to prevent a repetition of the catastrophic Paris crash in July 2000, and a £14m internal refit, Concorde could no longer earn its keep. At the end of October, after more than thirty years' service, the world's most glamorous aeroplane landed from its last commercial flight and yet another emblem of British genius taxied towards the museum.

The era of *le bang sonique* is now as remote as the age of steam, more in tune with the technological adventurism of the nineteenth century than with the accountancy-run, pocket-size pragmatism of the twenty-first. One prominent British engineer, asked to nominate his favourite structure built in Britain since 1945, unhesitatingly chose Skylon – the short-lived, cigar-shaped steel tower, more sculpture than building, that was erected on London's South Bank for the Festival of Britain in 1951. Even after an absence of more than fifty years, it still grips the imagination. So, even more emphatically, does Joseph Paxton's miraculous Crystal Palace – built in Hyde Park for the Great Exhibition of 1851, removed a year later to the area of south London that still bears its name, and subsequently destroyed by fire in 1936. Even in the minds of people who know it only from pictures, it stands as a memorial to lost momentum: possibly our last truly world-class public building. We have lost even our fabled ability to mark time. All we got for the millennium was the embarrassment of the Dome, a couple of breathtaking Underground stations on London's new Jubilee Line, elegant new footbridges across the Thames and the Tyne, an (admittedly stupendous) glass roof for the British Museum courtyard, the Eden Project and – against opposition (almost always a good sign) – the glorious London Eye.

Even buying a Rolls-Royce now looks more like an ironic,

post-Imperial gesture than a real endorsement of British superiority – a two-fingered salute to the new Germanic ideal of pared-down mechanical efficiency. Who, now, could name a famous British engineer? Where are the modern counterparts of George and Robert Stephenson, Thomas Telford and the Brunels – miracle-workers whose gifts to the world were on an almost biblical scale, controlling the uncontrollable? Names like *Pride* and *Enterprise* still attach themselves to ferries and pleasure boats, but where now does genius leave its mark on the national fabric?

Part of the problem is our nannyfied, safety-first attitude to risk – better be bland than sorry. During the drilling of the world's first underwater tunnel, beneath the Thames between Rotherhithe and Wapping in 1828, Isambard Kingdom Brunel had to swim for his life when water burst into the workings and two men were drowned. Such journeys into the unknown were commercially driven – no one in those days was competing for design awards – but it was accepted that risks, physical as well as financial, were as unavoidable as the navvies' blisters. The possibility of failure was admitted and accepted. 'No one in this country is allowed to fail now,' says the architect and author Neil Parkyn. 'You just miss certain targets.' The result is the same inversion of priorities, and negation of excellence, that infects the NHS, the universities, the broadcasting organizations and just about every city and town.

Bad buildings, like bad television programmes, have a lot going for them. Their architects have dim or penny-pinching clients abetted by wall-eyed, risk-averse local planners and their committees – timid, mug-of-Horlicks mediocrities clinging to their comfort blankets, afraid of being taken expensively to public inquiry. With low- and medium-rise redbrick rubbish, sheds in all but name, these serial killers have throttled one historic town centre after another. And you have the government. A few years back, I was one of two writers invited by the Department of Culture, Media and Sport to help an ad hoc committee put into words a new code of practice intended to improve the design of public buildings funded by the Private Finance Initiative (PFI). The committee was composed not just of effete architectural creatives in silly spectacles, but included burly, cut-the-crap hard men from the construction trade. All saw buildings potentially as objects of beauty, which they regarded as self-evidently desirable. Not so, however, the representatives of HM Treasury. Their habitual response to loose talk of 'beauty' was: how do you define it? How does it make the building cheaper? The project staggered on for a few weeks, like a ballerina humping a sandbag, then collapsed and was left to die.

In a sense, we should be glad to see the occasional heroic architectural stinker. It is evidence at least that adventure and innovation are not dead, that bold statements are permissible and that the possibility of excellence, too, remains alive. Without the risk of a Tricorn Centre you get no Lloyds building, no Pompidou Centre, no Gateshead Millennium Bridge. But bold design depends on enlightened patronage. With a few notable exceptions (the new stations on the Jubilee Line, for example), this has gone the way of stovepipe hats, watch-chains and mutton-chop whiskers. The Millennium Dome was less an architectural or constructional failure than a political one – a wonderful structure with a hopeless client.

Everywhere the PFI mentality prevails: building to a budget, using standard design ingredients and with a selection process weighted in favour of the lowest bidder. Design in public buildings is biased towards functionality, not aesthetics, and – though most schools, hospitals and prisons will do more or less what it says on the tin – the chance of a best-in-the-world structure is only marginally higher than that of Norwich City winning the European Champions' League. The risks are not of the kind understood by Brunel and Telford. They are not the gung-ho hazards of the buccaneer but the stifling constraints of inertia; of infrastructure gagging for investment; of small men despoiling the work of giants. The railway is like an Old Master peeled of its varnish and left out in the rain. London Underground – in its day, another world-class example of technological and commercial vision, driven by sheer sod-the-risks nerve – now stands as a metaphor for hell.

You could argue that the entire modern world, global economy and all, came off Britain's drawing boards in the nineteenth century. The engineering expertise that built railways, bridges, water and sewerage systems and canals was the country's most visible export. By the time of his death in 1871, a single British contractor, Thomas Brassey of Chester, had built one eighth of the world's entire railway mileage. The waves, too, were ruled as never before or since. For more than a century Britain's ocean liners dominated the North Atlantic. In particular, the Clydebank shipyard of John Brown & Co. produced two of the most iconic vessels in the entire history of passenger transport. Cunard's *Queen Mary* (launched 1934) and *Queen Elizabeth* (1938) were global stars whose celebrity ranked alongside the royal personages whose names they bore. Simply to cross the Atlantic in one of these was a celebration of British supremacy; an experience that, like Concorde later, would enrol the traveller in a high-gloss global elite. Change, when it came, was elemental. By the time *Queen Mary* claimed the Blue Riband in 1938 with a 30-knot Atlantic crossing in three days, 21 hours and 48 minutes, she was already facing her aerial nemesis. Eleven years earlier, Charles Lindbergh's *Spirit of St Louis* had completed the trip in just 33 hours.

For glamour, then, we looked to the skies. In the Sunderland, Lancaster and Spitfire we'd had the Second World War's most famous flying boat, bomber and fighter. In the de Havilland Comet we gave the world its first jet airliner (though in its chapter of accidents and early grounding, its chief gift to posterity was an improved understanding of metal fatigue). Now, as the great queens sank into memory, a new icon blazed around the globe, turning heads wherever it went. Even more than the queens, Concorde fired the imagination, and won the pride of people who had no hope of ever riding in it. Matching the grace of a swan with the speed of a missile, it delivered in their distilled essence all the elements of the national self-image – beauty, power, enterprise, courage. It was also, as it would turn out, a monument of hubris, the dawn and twilight of supersonic travel combined in the same act of commercial miscalculation. In Concorde's very exclusivity, unintended result of its failure to win orders from foreign airlines, lay the seeds of its own demise.

On the railways, British supremacy died with steam. We made some decent diesels (the 1955 Deltic prototype, driven by engines adapted from warships, was briefly the most powerful diesel-electric in the world), but – with intensifying competition from France, the US and Italy – they could not hope to match the output of the imperial past, when engines were exported in their tens of thousands from Glasgow,

Manchester, Leeds, Newcastle and Bristol. In the age of electricity there was a late bid for technological leadership in the APT, or advanced passenger train, a truly astonishing piece of high-tech engineering that neutralized centrifugal force by tilting the train on bends. For a while, British Rail's engineers at Derby were world leaders in tilt technology, but problems with the prototype eroded political and commercial will, and in the mid 1980s, after seven years of development, the APT was shunted off into the sunset and the UK locomotive industry all but disappeared into the archives of steam preservation societies and the National Railway Museum. Twenty years later, tilting trains at last are being run by Virgin on the west coast main line between London, Manchester and Preston, but the 140mph, Fiat-designed Pendolinos are manufactured at Savigliano, near Turin, by the French transport giant Alstom, and trucked in kit form to Washwood Heath in Birmingham, where British engineers at least have the job of assembling them.

Engineering genius is a tender plant when exposed to long winters of neglect. The 10,255 miles (16,500km) of UK rail routes use a total of 19,733 miles (31,750km) of track. To keep it all safe as it wears, it needs to be replaced at a rate of 500 miles a year. This was, more or less, what British Rail used to achieve before privatization. In the dog days under Railtrack, it fell to as little as 150 miles, so that when Network Rail took over responsibility in 2002, it inherited a backlog of 4,000 miles of track – ballast, sleepers, rails, the lot – already overdue for renewal. By late 2003 Network Rail was achieving 700 miles a year, and by early 2005 it had improved to 900, accelerating the annual catch-up rate from 200 to 400 miles. Even so, it would be several more years before the system was back in equilibrium, and the consequence in the meantime would be more patch-up work on old track, more delay, more bus transfers, more slow running.

If there is little inspiration from the railways, there is not much more from the runways either. Airlines flying into some of the world's most congested airspace not unreasonably want more tarmac. People living near Stansted and Heathrow plead for their last few shreds of peace. It is by the incessant drip of piecemeal expansion that lives and landscapes are degraded. What we are crying out for, said Richard Haryott, chairman of the Ove Arup Foundation, is not a new runway at Stansted but 'a great soaring piece of imagination. To me it shrieks out that we need a new 24-hours-a-day, seven-days-a-week airport in the south of England. It would have to be over water, probably on an artificial island in the Thames estuary. But we don't have the will to do it. We deem it not to be nationally worthwhile. We would rather spend it on prisons.'

Big ideas shrivel as they pass from desk to desk. The idea of Crossrail – running trains on main lines *through* London instead of just into it – was first put forward in 1906. We are still talking about it, and still waiting. We're still waiting, too, for a solution to London's traffic chaos. The concept of the orbital motorway – what a first that would have been! – got its earliest airing in 1905. In the meantime there have been plans for a five-ring orbital system (in 1944) and a three-ring system including an inner-city motorway (1969), grand designs that were transmuted by timidity into the atrophied horror of the M25 and the fly-whisk solution of the central London congestion charge. We did get Stansted airport, but only at the expense of a more ambitious scheme for Maplin that would also have given London a new container port and staunched the haemorrhage of trade to Antwerp and Rotterdam. Twenty years later, the most exciting plan to hit the capital since the Crystal Palace – Richard Rogers's scheme to enclose the South Bank arts complex in a huge glass dome – found only moths in the Arts Council's wallet.

Other bits of the national infrastructure meanwhile continue to waste away. The price rises agreed by Ofwat may ease some of the strain on ancient sewers and water mains, but other questions, no less urgent, are pressing for answers. No matter how electricity is generated in future, there remains the problem of piping it into our homes. Like the water system and the railways, the pylons and cables of the National Grid are approaching the end of their operational lives. 'In the next ten or fifteen years,' said the Institution of Civil Engineers' David Anderson in 2003, 'much of the infrastructure will need replacing.' This is trickier than it sounds. You can't turn off bits of the grid for months on end while you rewire it – this would bring down the government. To keep the supply running you often have to erect brand new lines, sometimes over completely new routes, before the old ones are disconnected. And new routes mean protracted wrangles with landowners and bitter arguments with objectors, who inevitably want unsightly cabling laid underground, at between twelve and twenty times the cost of the £400,000 per km it takes to rig it overhead. 'Although building the line itself might take only two years,' he said, 'the process from concept to commissioning can take ten.' Which, by simple arithmetic, leads one unerringly to the conclusion that someone had better do something soon.

The House of Commons Select Committee on Trade and Industry – the parliamentary watchdog that oversees the DTI – seems to think so too. Alarmed by major power failures in the

US and Italy, and by short but serious blackouts in central London and the West Midlands, it took a hard look at the national electricity network. Its report in early March 2004 spelled out the problem in its usual no-frills language. There was, it said, 'a danger that there is currently insufficient investment in the network to replace in a planned and orderly way equipment which is reaching the end of its life'. Behind it lay the same old problems that had wrecked the railways and threatened water – 'pressure to minimise operational expenditure'.

> While this pressure has doubtless resulted in reducing some inefficiencies, we think that to continue it may be counterproductive for network performance: ageing assets are likely to require more, and more skilled, maintenance. In relation to this, we also raise concerns about a likely shortage of highly skilled staff both to maintain and to reconstruct the network . . . We were told that, simply to maintain present performance levels, capital expenditure by the network owners would have to double.

The political message was equally clear. Bigger electricity bills might generate hostile headlines, but a failed system would be a catastrophe. 'We consider that the Regulator's concern to reduce costs to consumers should now be tempered by a greater emphasis on ensuring that electricity network owners have the financial resources necessary to secure a viable long-term electricity supply.'

Bills up, or balls-up. That seemed to be the choice, though the DTI was not falling over itself to express a preference. Its response was so evasive it might have come straight from a minister on the *Today* programme. 'In fulfilling its statutory duty to protect the interests of both current and future consumers, [the Regulator] Ofgem seeks to strike a balance to ensure that prices are set at a level which provides value for money both in terms of current costs and funding for maintenance and future investment.' So there. We're in good hands.

The degraded remains of a golden industrial past are a new kind of junk, but not the newest. For that, we have to look upwards. There is no official record of how much scrap is now orbiting the planet, but most authoritative estimates given by official sources are traceable to the California-based Center for Orbital and Reentry Debris Studies (CORDS), which has been working on the problem, and seeking its solution, since 1997. There is all sorts of stuff up there: spent launch vehicles, dead satellites and all manner of leftovers from what CORDS calls 'deployments and separations'. It lists, for example, explosive separation bolts, lens caps, momentum flywheels, nuclear reactor cores, clamp bands, auxiliary motors, launch vehicle fairings and adapter shrouds. And that is not all. What with atomic oxygen, solar heating and solar radiation, it's a hostile environment up there. Things degrade and flake apart, creating a flying storm of tiny solid particles made up of paint flakes and flecks of 'multilayer insulation'. Solid rocket motors also produce debris from motor casings, aluminium oxide exhaust fuel particles, nozzle slag, 'motor-liner residuals', solid-fuel fragments and 'exhaust cone bits'. As objects break apart, so the fragments collide with each other and multiply in number as they diminish in size, thus creating dense and dangerous 'debris clouds'. Some of the break-ups are the result of explosions caused, apparently, by the coming together of propellants and oxidizers. Another hazard is the shower of frozen nuclear reactor coolant leaking from old Russian Radar Ocean Reconnaissance Satellites (RORSATS), designed to monitor foreign navies, of which thirty-one were launched between 1967 and 1988. NASA has calculated that there are more than 70,000 frozen 2cm droplets at altitudes between 850 and 1,000km.

The risks for anything attempting to run this space gauntlet are spectacular. Even a fleck of paint travelling at 25,000mph can strike with the force of a bullet. Space shuttle windows often have had to be replaced because of paint damage (CORDS shows a photograph of a windshield with a 4mm diameter crater in it, caused by a flying fleck of white paint no more than 0.22mm in diameter), and shuttles frequently have had to change course to avoid larger objects. In low earth orbit – i.e., below 2,000km – the average impact speed is 10km/sec, or 21,600mph. Ballistics experts say a bit of aluminium only 1.3mm in diameter would have the penetrative power of a rifle bullet. A 1cm piece would have the destructive power of 'a 400lb safe travelling at 60mph'. A piece 10cm long would do the work of twenty-five sticks of dynamite.

Ideas for limiting space junk in future include 'tethering' or 'catching' bits of spent equipment – lens caps, separation bolts – that previously would have been jettisoned, ensuring that unused fuel is burnt off, and parking scrap spacecraft in designated graveyard orbits. But, although domed heads are forever being scratched, the cleansing of space is as remote a possibility as the cleansing of the world's oceans – another consequence of mankind's mad belief in the absorbent power of infinity. There is so much big talk about the future of space – tourism, colonization, industrialization – and so much interest from privateers that it is not at all easy to separate the crazies from the serious scientists. Some of the maddest talk has come from people authorized to speak on behalf of their national governments.

Theoretically, almost anything is feasible. You could put a branch of Tesco in orbit around Mars, or build on the moon a city the size of Los Angeles. Feasibility, however, has a habit of overreaching itself. If something *can* happen, then it *will* happen is the way the thinking goes – the classic punter's fallacy, and the reason you never meet a bookmaker with holes in his shoes. In October 2000 an influential member of the Chinese Academy of Sciences, chief scientist of the country's moon programme Ouyang Ziyuan, spelled out his thoughts in a collection of essays, *Academicians Envisioning the 21st Century*. What Ouyang himself envisioned was more like the plot-line of a space cartoon than the desiccated Maoist mission statement suggested by the title. By 2005, he said, Chinese astronauts would have begun the work of converting the moon into a naturally orbiting space station generating its own electricity. By 2010 the base would be able to support teams of scientists for weeks on end. By 2015 it would have a permanent population, well advanced in developing the factories and farms necessary to sustain the self-sufficient village – first stage of the eventual moon city – that would be established by 2020. Lunar industries would exploit metals and gases mined from the moon rock, and surplus energy from solar-powered generating plants would be sent back to earth.

One recalls the words of Sir Richard Woolley on his appointment as Astronomer Royal in 1956. The idea of space exploration, he declared, was 'utter bilge'. He did not mean that a moon landing was technically unfeasible; merely that it was pointless. Even as Apollo 11 was approaching its giant leap for mankind in July 1969, he refused to come out from behind his plain man's barricade of common sense. The moon landing, he said, was 'not only bilge but a waste of money'.

He has been much mocked, but history has not exactly rammed the words back down his throat. The cost of the eight-year Apollo programme was colossal – equal to the United States' entire expenditure during the same period on lipstick and potato crisps – and it was the sheer, superhero scale of the human adventure, the 'small step for man' not the scientific discoveries, that caused jaws to drop. 'We now know,' said NASA, 'that the moon is made of rocky material that has been variously melted, erupted through volcanoes and crushed by meteorite impacts.' The moon had no atmosphere; no living organisms or fossils; the thing was, and always had been, inorganic. The six Apollo missions between them brought back piles of geological detail (thickness of crust, age of rocks, etc.), and 382kg of samples to put under the microscope. But that was basically it. All very beard-and-anorak, and all not very Dan Dare. The moon's last human footprint was left in December 1972, so long ago that most of the mission crew are counting their descent into old age, and first-hand experience of manned lunar flight will soon pass into eclipse.

This did not deter Ouyang Ziyuan. In May 2002 he launched Beijing's national science and technology week with an official, on-the-record reiteration of his earlier promise: 'China is expected to complete its first exploration of the moon in 2010 and will establish a base on the moon as we did at the South Pole and North Pole.' A Chinese rabbit, dog, monkey and snails soon would orbit the globe aboard the spacecraft Shenzhou II. Men would follow, on schedule, in 2005. Twelve were already in training.

In the meantime the Soviet Union remained the only country other than the US to have fired humans into orbit, though post-Soviet Russia has cracked under the financial strain and is even further out of the imperial space race than its former Cold War enemy. On the western side, all the big talk now comes not from governments but from corporations – many of them so small that you would doubt their ability to run a company Vauxhall, never mind a space fleet. There is talk of private space stations and moon-orbiting cruise liners, all within the usual 'fifteen or twenty years'. For American would-be adventurers (and the US is their heartland), the frustration is that you can't get anything off the ground without the permission of the Federal Aviation Administration (FAA), and if it's a space launch you will need also the help of NASA, an agency whose attitude was satirized by one entrepreneur as No Access to Space for Americans. The fact is that governments have never been keen on privateers. Despite all attempts by the boldly named United Nations Office for Outer Space Affairs to establish a canon of space law, the moon lies far beyond the scope of any conceivable police force and its legal status remains untested. When Boeing tried to buy back leftover parts of the Saturn rocket for its own use after the Apollo programme folded, it met with little encouragement. Governments worry about the Bond-like spectre of undesirables in orbit, and about who would control the companies if they got there. It could be like Captain Kidd in the Caribbean. And what could you do if one national group on the moon got into a fight with another?

But still the talk goes on. Increasingly outside China there is a feeling among space professionals that the key to progress is privatization, that privatization depends on profit, and that the way to profit is tourism. Technically (of course) it's all perfectly feasible, though financially it's dancing in the dark. If manned space flight in the past had had to pay its way, there would have been no Apollo missions and no International Space Station. If it is to have a future, it will have to 'wipe its face', as the money men say. And this is where the fifteen-year crystal-gazers go into hyperdrive. Sooner rather than later, they say, one of the many reusable flight systems currently on the drawing board will find a commercial sponsor and space will be up and running. Tourism will come first; then colonization; then industry. 'All it needs,' says one UK engineer with a range of designs ready to take wing, 'is a critical mass of people taking it seriously.'

What we are asked to imagine is a space bus shuttling passengers to a space station in low earth orbit, where it would connect with onward services to the moon. By some accounts the lunar transporters would not only be operated from the moon itself, they would be *built* there. (Yes, I know, but stay with me.) The visionaries – and they are nothing if not that – see the moon, much like the plains of America but without the inconvenience of native peoples, being colonized by stages. The pioneers would not be men and women at all, but robots – skilled electronic craftsmen with fully articulated hands that could be worked from earth with great accuracy, making use of sophisticated and powerful machinery. They could put up solar panels to generate power, and, exactly like human frontiersmen, build with local materials. After that would come manned stations, space hotels and the industrial revolution. All the world's population and pollution problems would be solved by shifting its heavy manufacturing industry into deep space where it could be serviced from the moon.

In 2002, the British Interplanetary Society's lunar spokesman, Richard Taylor, put it to me like this: 'You could manufacture anything there, using materials found on the moon or mined from asteroids. Some asteroids are largely metallic and contain easily recoverable rare metals such as platinum and iridium. One small asteroid would contain more platinum than we have ever had on earth.'

There is also plenty of iron. Operationally, with low gravity and no lunar atmosphere, launching a vehicle from the moon would require much less velocity than launching from the earth, which means substantial reductions in both technical difficulty and cost. According to Taylor, a gentle lift-off from the lunar surface would cost only 5 per cent of an equivalent blast-off from earth – which has implications not only for local tourism but for onward passage to Mars. It is difficult not to be sceptical.

One voice to which the government might be expected to listen is that of QinetiQ, the government-owned science and technology crack-squad formerly known as DERA, the Defence Evaluation Research Agency. Its space consultant, Richard Crowther, took the lunar colony idea with a large pinch of crazy dust. He pointed to the orbiting international space station (ISS) being assembled by a US-led governmental consortium including Russia, Canada, Japan and ten members of the European Space Agency. There was no way, he said, that such a venture could pay its way commercially, which raised serious doubts about the whole idea of private enterprise in space. Surprisingly, however, he did not reject Taylor's vision of an extraterrestrial industrial estate, though he saw it strictly in terms of automation – robot workers managed by computers, with no need of a human colony. 'It's much more efficient to rely on machines,' he says. 'Unlike humans, they don't have to be heavily protected from hazardous environments.' At most, it might be necessary for a small number of technicians to share the task of manning a lunar service station. When might all this happen? 'There could be a space station on the moon within twenty years. But industry in space? Not in my lifetime [in 2002 he was forty-one].'

Space tourism didn't impress him either. 'Passengers can't just get into a spacecraft and blast off. You have to spend months training to be able to deal with all the eventualities that might occur.' There is an Alton Towers dimension to zero-gravity training that some people might actually enjoy, though the enjoyment might pall when they make their first attempt to use a zero-gravity lavatory, with or without zero-gravity constipation. So who knows? How deep, how soon, how injurious, will be man's imprint at the last frontier? Already we know part of the answer. More than a million bits of waste have coalesced into a whirling junk-blanket around the planet, with some 1,900 tonnes in low earth orbit. Whatever this indicates, it is not lessons well learned. Space is time as well as distance. Will we now launch ourselves backwards into the psyche of the troglodyte, looking for holes in the universe? With more and more communications, surveillance and location satellites nudging into the traffic and onward into the celestial junkyard, are we looking to a policy of skyfill?

The nearest thing to a certainty is that, for better or worse, the next breakthrough in space transportation will not come from the United Kingdom, and probably not from anywhere in

Europe. The European Space Agency has declared that 'man's return to the moon is not for the immediate future', though to the chagrin of pioneers who believe they have already answered the question, it has begun yet another programme to 'investigate concepts and technologies' for a reusable launch vehicle. The UK government meanwhile grandly holds itself aloof. 'It is not essential to have launchers as part of a dynamic space policy' is its somewhat mystifying policy position.

That we even have a national space agency, the British National Space Centre (BNSC), is news to most people. Not only this: we actually have a minister for space – Lord Sainsbury of Turville, former head of the grocery chain and now parliamentary under-secretary of state for science. Fewer people know that than know the atomic number of argon. The BNSC is very good at keeping lists of UK companies involved with communications satellites, and very good at promoting exports, but men on the moon are as far from its agenda as free tickets to the Martian Olympics. Just to be on the safe side and stifle any fancy ideas, we are closing university chemistry and physics departments.

The robots, however, are not giving up. They may not get to the moon, but they are hell-bent on inheriting the earth. More and more, the basic transactions of our daily lives involve dehumanized, and dehumanizing, interactions with electronic circuitry in a progressive disassembly of personal relationships. Trying to talk to your bank, telephone or internet service provider, train operator, credit card or electricity company is to be sucked into a cat's cradle of multiple-choice, digital loops that directs you in unbreakable circuits of rage from one dead end to another. If you do get through to a real person, it is only to be told to ring a different number. The one human voice you *can* still depend upon is that of the station announcer, for whom lateness has become almost a force of nature – like snow and leaves on the line, needing no explanation beyond the miracle of its own existence. 'We apologize for the late running of this service. This is due to the late arrival of the incoming train.'

But I am ranting. The voice in my head shouts in furious, self-harming impotence like that of an old man who can't work the video. Yes: part of the problem is that things aren't what they used to be. Modern life is rubbish. But the other part, and by far the bigger, is that things stay exactly the same. Intellectually we may be conservationists, recyclers, sustainable developers. Instinctively we are driven by our appetites. The caveman inside us will not go away.

THE END . . .

Or perhaps not quite. We are *homo sapiens*. Although, like other species, we are driven by our native urges, we have the intelligence to understand cause and effect, and to accept responsibility for the consequences of what we do. We know the earth is not finite. We know we cannot go on spending its resources and depleting its energy without risk to our own as well as future generations. As cavemen we are programmed to preserve our genetic stock, but this does not just mean taking flight from wolves. Just as importantly, it means recoiling from our own baser instincts when they conflict with our long-term needs. Yes, as individuals, there is much that we can and should do. In towns, villages and countryside we still see the fruits of our own inherited genius. The 'environment' of which they are a part is not some theoretical concept in the minds of dreamers. It is the entire fabric of our lives, dawn to dusk, birth to death, generation to generation. Let us protect it for all it is worth. Press our leaders to look further than their elected terms of office, and live our own lives as we know we should. Do all the obvious things. Conserve water and energy. Recycle. Take our rubbish home. Explain to doubters why the green wheelie bin matters. Support enterprises that recognize their responsibility to conserve, and avoid those that don't. Whatever Ouyang Ziyuan may say, it's the only planet we'll ever have.

Afterword

Forgive me while I waffle. At my request, the publisher agreed to hold back these pages at the end of the book so that I could add any developments that post-dated the completed text. It is mid March 2005, the deadline is upon me and I'm panicking. What is there to say?

Defra at the turn of the year raised its petticoats and offered a flash of green garters. The secretary of state, Margaret Beckett, 'unveiled' a high-tech recycling plant at the Winchester branch of Tesco, which – like the NEWS plant described in Chapter 8 – uses infra-red scanning to sort plastic, glass and aluminium. If the trial is successful, and if 'customer feedback' is positive, the technology will be 'rolled out' to other Tesco branches across the country. I take this to be good news. A couple of days later, Mrs Beckett announced that Wrap would spearhead a £1.2m project 'to help retailers pilot new ways of encouraging householders to recycle their waste at supermarkets'. Defra would also be working with local authorities to 'explore original ways to motivate householders to recycle'. What this means in practice, we'll have to wait and see.

Otherwise the department's principal product continues to be talk, albeit with multiple calls for action from others. Its addiction to obfuscatory New Labour jargon remains unbreakable, making it hard for anyone outside the spinning circle to know whether to applaud or (as Philip Larkin might have put it) vomit into the nearest bowler hat. In early December 2004 it announced its 'first call' for 'waste research proposals' to support its three-year 'waste and resources research programme', which, it said, 'aims to provide a sound evidence base for policy development, implementation, monitoring and evaluation for sustainable waste management at both the national and local levels, and to aid the introduction of innovative solutions within the UK'. In case this might be thought in any way unclear, it set out six specific themes for applicants to ponder – 'sustainable resource consumption and management', 'systems for resource recovery', 'social dimension', 'environment and health', 'economics' and (my favourite) 'decision tools'. The bowler hat beckons.

On the very same day, the environment minister Elliot Morley 'unveiled' four new consultation papers. According to Defra itself, these would:

- underline the need for effective integration between the development of municipal waste management strategies and planning processes and policies regionally and locally, with full appraisal of options in their development;
- provide the framework for setting a long-term vision for fifteen–twenty years ahead and a short-term detailed plan of action;
- emphasize the need for early and continuous community engagement.

Coming on top of such a ponderous, belated statement of the blatantly obvious, it's that 'early' that makes it so special. I enjoyed also Morley's Damascene conversion on the road to common sense. 'Not collecting or managing waste,' he declared on 6 December 2004, 'is not an option.'

His colleague Alun Michael was at it too, heading off to help clean up a 'litter hotspot' in Cardiff. His broader purpose seems to have been to draw attention to the government's Clean Neighbours and Environment Bill, which it set out a few days later. This had a promising look to it. Ideas included: fixed penalty fines for litter, graffiti, fly-posting and 'dog offences'; new powers for local authorities to remove abandoned cars, force businesses and individuals to clear litter from their land, and businesses to clear up their own street litter; confirmation that cigarette butts and chewing gum are legally litter. There were also stronger measures, including increased fines, to attack fly-tipping, and powers for local authorities to reclaim from supermarkets the cost of collecting abandoned trolleys. All this, though long overdue, is laudable, and I hope everyone will share my pleasure in the proposal to 'extend the list of statutory nuisances to include light pollution and nuisance for insects'.

Also in December 2004, Elliot Morley announced that local authority recycling targets for 2005–6 would be capped at 30 per cent, provoking Friends of the Earth to complain of 'a breathtaking U-turn' and 'a kick in the teeth' for green-minded local authorities such as Daventry that were already achieving 40 per cent or more. Even the Local Authority Recycling Advisory Committee (LARAC), which broadly welcomed the cap, warned that it might cause high-performing councils to 'take their foot off the pedal'. At the same time, Defra revealed that England was 'set to launch the world's first allowance trading scheme for municipal waste'. The idea, it said, was to make life easier for councils – 'working within a challenging efficiency agenda' – to meet their landfill reduction targets. From April 2005 they would be set allowances for the amount of biodegradable waste they could legally landfill, but would be permitted to trade allowances between themselves. Those already using recycling plants would be able to sell their spare landfill allowances to others that had fallen short.

This does not guarantee the rescue from landfill of a single extra potato peeling or scrap of paper. What it does is hoist the weak on to the backs of the strong, and massage the national bottom line to conform with minimum annual targets. Morley said it himself: 'While the targets are challenging, [the Landfill Allowance Trading Scheme] is an innovative approach which gives authorities the flexibility to decide how and when to make the necessary changes in the way they handle their waste, while ensuring that England meets national and international obligations in the most cost-effective way.'

It is neither stick nor carrot. Penalties (at £150 per tonne) will be imposed only on those that don't buy enough allowances to cover their landfill total. But how were they supposed to plan, or fix their budgets, in the face of such uncertainty? There was no guarantee that enough spare capacity would be available for them to buy, and no mechanism other than market forces to control the price. Peter Gerstrom, of the Institution of Civil Engineers, thought the scheme would be of genuine help to smaller local authorities unable to afford their own recycling plants, and that the threat of fines would 'frighten local authorities into taking their responsibilities seriously'. But, he said: 'The whole thing is not transparent.

It doesn't work if you haven't got transparency.' Among the rumours circulating in March 2005, just three weeks before the scheme was due to come into force, was that at least one council was offering spare allowances at £32 a tonne. It's a funny way to save a planet. In the opinion of Biffa's Peter Jones, it was 'sweeties for the good guys – a transfer tax from non-acting Labour councils to proactive Lib Dems and Tories'. To add to the inequality, the scheme has not been adopted in Wales, which, in the words of another industry insider, 'will mean everybody will be equally bad'. Morley did, however, hint that English backsliders would not be allowed to fail for ever. On 4 March he announced that he would be 'turning the spotlight on', and 'personally engaging with', the persistently poor performers.

Four days later, the successful accomplishment of the UK's modest 17 per cent national recycling and composting target for 2003–4 was paraded in triumph. Once again the UK had performed entirely to its own satisfaction, though the Local Government Association warned it was 'extremely unlikely' that the country would meet its landfill target for 2010 (a reduction to 75 per cent of the total for 1995). Failure, it reminded the government, would result in fines of up to £180m a year, or £500,000 a day. There were other scraps of good news. The overall quantity of municipal waste went down by 1 per cent, from 29.4m tonnes in 2002–3 to 29.1m in 2003–4, but even this had a measure of cold water poured over it. LARAC feared it might have been 'just one year of good fortune', brought about by 2003's exceptionally dry summer, which had substantially, but only temporarily, reduced the weight of green waste.

More positively, there were reductions in the overall amount of waste sent to landfill, from 22.1m to 20.9m tonnes, and in the amount of garbage collected per head from 520 to 510kg. Part of this may have been accounted for by the increased volume of recyclable material taken to civic amenity sites and 'bring banks' – 58 per cent of recyclable material (2.6m tonnes in 2003–4) now follows this route. Recycling rates in England varied greatly from region to region. Despite recycling 80 per cent more waste in 2003–4 than it had done in the previous year, the North East still managed an overall recycling rate of just 11.9 per cent. Champion regions were the East, at 23.2 per cent, and South East at 22.7 per cent. The national target for 2005–6 is 25 per cent.

The industry's clamour for clarity on Waste Acceptance Criteria for hazardous material (see pp. 245, 246) was answered on 15 December 2004 by a consultation paper that set a deadline of 7 March 2005 for responses – just four months before the criteria would have to be enforced. On 2 March, Elliot Morley added to the gaiety of the season by urging businesses to be ready for the changes in July. Failure to comply, he warned, 'could leave them out of pocket or in jail'. Old hands reached for their usual pinch of salt. The insiders' view was that the proposed criteria were not only too late but also unrealistic, having been prepared by consultants from too small a sampling base. Jail or no jail, there was little hope that they could be achieved in time.

Meanwhile there remained deep uncertainty over the whereabouts of the hazardous waste that had apparently dipped below the radar since July 2004. The Environment Agency line was that companies were now treating it 'on site' rather than shipping it to landfill, but those in the business were more inclined to believe it was being stockpiled. Such hazardous waste that *was* on the move had to make prodigious journeys by road – from South Wales to North East England, for example. Whatever the architects of the Landfill Directive intended to achieve, it cannot have been this.

Anyone hoping for a swift resolution of the uncertainties over the WEEE Directive (see Chapter 7) had also been frustrated. The DTI had held the inevitable consultation, but by 9 March there was still no final recommendation on the proposed National Clearing House, and thus no clarity on how defunct electrical appliances would be collected and allocated to producers after the directive came into force in August. The DTI said an action 'timeline' would be announced 'in the next few days', but sources within the industry were already suggesting that the clearing house was a dead duck, the timetable was 'impossible' and – following the well-trodden route from inertia to farce – the UK would have to postpone implementation of the directive at least until early 2006.

There was foot-dragging, too, at the Office of Fair Trading. Its compliance audit of the Supermarket Code of Practice (see Chapter 4), promised initially by the end of 2004 and then delayed until 'early 2005', had still not been published. 'Next few weeks' remained as close as it could get to a date. Meanwhile the City forecast annual profits for Tesco in excess of £2 billion.

The accident-prone Office of the Deputy Prime Minister was facing further setbacks in its ambition to build homes on recycled industrial land. The European Court of Justice ruled that previously contaminated sites would have to be reclassified as 'landfills', triggering what the property consultants Savills described as a 'whole range of legal and financial obligations' that could be fatal to development plans – a problem that came on top of the existing uncertainties about the transport and treatment of 'hazardous' building and demolition wastes. All this will only intensify developers' appetite for trouble-free, greenfield sites in the countryside.

The sea, too, was taking a pounding. The Worldwide Fund for Nature (WWF) in its Marine Health Check 2005 recorded severe depletions in UK waters of everything from dolphins to maerl beds, almost all of it due to overfishing and destruction of marine habitats by bottom trawls. An attempt to count the surviving population of the common skate had collapsed when researchers failed to find a single fish. The WWF quite reasonably spoke of 'seas in crisis'.

Fish were not the only things at risk. In March, the *Sunday Times* published evidence from a former safety officer that there had been persistent breaches of safety in and around the Dounreay nuclear plant in Caithness. He claimed that high-level radioactive waste had been washed down drains designed for low-level waste; that effluent samples for analysis were collected in a wellington boot on a string because sampling machinery was 'a heap of rust'; that the discovery of radioactive waste on the neighbouring public beach was 'covered up'; that radioactive containers in dumps were not properly marked.

Unveiling her green paper on the future of the BBC, the culture secretary, Tessa Jowell, guaranteed the licence fee until 2016, but sought to exact a price. In return, the BBC would be expected to produce more and better 'landmark' programming, stop 'chasing viewers' with lowest-common-denominator format shows and wean itself off its nourishment-lite diet of 'reality' programmes, makeover shows, repeats and American imports. According to the Department of Culture, Media and Sport, there was a 'marked perception' in focus groups that the

BBC's standards were falling, especially in television. Rubbish in future was to be the exclusive province of commercial channels, which should be left to get on with it unopposed. The BBC, said the green paper, should 'stay out of bidding wars for expensive foreign imports except where it is clear that no other terrestrial broadcaster would show all the programmes or films in question, or that the acquisition would clearly contribute to a public purpose'. Against all probability, the one place where rubbish seemed to be in retreat was the BBC.

Index

Also from Eden Project Books

The Origin of Plants:
the people and plants that have shaped
Britain's garden history

by Maggie Campbell-Culver

From a mere few hundred indigenous plants in 1000, Britain now boasts the widest range of flora of any nation on earth. *The Origin of Plants* tells the fascinating story of how and why Britain's gardens and private and public collections came to be spectacular. Europe and the Near East, Russia and North America, South America and South Africa, India and the Antipodes, Japan and China, have all yielded a remarkable bounty to warriors, explorers and plant hunters.

Critically acclaimed, rich in colourful detail and anecdote, Maggie Campbell-Culver's work is also timely. With some of the world's plants under extreme threat in their native territories, Britain's plant collections have become of crucial importance. This is a must-have book for all plant lovers.

'A most welcome and accessible reference work.'
Times Literary Supplement

'Full of facts and legends, this will appeal to both the general reader and the more difficult-to-please scholars.'
Penelope Hobhouse, *Gardens Illustrated*

ISBN 1903 919401

Paperback non-fiction

£9.99

A Good Life:
the guide to ethical living
by Leo Hickman

- Should I buy an organic apple from New Zealand, a fairtrade apple from South Africa, or a non-organic apple grown locally?

- What are my household cleaners doing to my family's health?

- Who deserves my charitable donations the most?

If we knew how to go about it, millions of us would like to live a more 'ethical' life. Our multiple-choice society encourages us to exercise our purchasing power to the max, yet the sheer range of ways we can spend, invest or give away our money throws up more dilemmas than it seems to solve. Addressing every area of our daily lives, from the food we eat to how we furnish and heat our homes, from how we spend our leisure time to how we look after our health, travel and financial arrangements, *A Good Life* examines the ethical dilemmas we face daily, acts as a positive guide through a maze of choices and their consequences, and provides an indispensable directory of goods, suppliers, companies and organizations to help us on our way.

ISBN 1903 919592

Paperback non-fiction

£15.00

A Life Stripped Bare:
tiptoeing through the ethical minefield
by Leo Hickman

Over the course of a year Leo Hickman conducts an unusual experiment . . .

Is it possible, in the twenty-first century, to lead a 'normal' life – to have a job, kids, a mortgage, holidays in the sun – but at the same time be respectful to the planet and the people who share it? Could Leo, his partner and baby live a more 'ethical' existence? Or is a 'good life' the preserve of fanatical eco-warriors, new-age spiritualists and the organic-product-endorsing Hollywood Set?

To find out, Leo invites three 'ethical auditors' into his home. As they explore the murkier recesses of his life – his bathroom, his fridge, his holiday plans, his shady DIY habits – he discovers that ethical living involves a lot more than changing his brand of washing-up liquid. And appealing for help and guidance on the internet, he is inundated with advice from the extreme: 'Having children is the most unethical thing you can ever do, so try ceasing that for starters' to the extraordinary: 'The most sustainable food sources are skips!' Innocents abroad, Leo and his family are in for some nasty shocks, but there are revelations too.

This is the record of an extraordinary transformation. Both funny and inspirational, it is a mine of information for all of us with a conscience. Who knows, it might change your life, too.

ISBN 1903 919606

Paperback non-fiction

£10.99

Dancing at the Dead Sea:
journey to the heart of environmental crisis
by Alanna Mitchell

Does the human race have a death wish? One hundred and fifty years after the publication of *On the Origin of Species*, award-winning writer Alanna Mitchell sets out on a journey to the world's hotspots – where the environment has been all but destroyed – to pick up where Darwin left off and examine not the origin but the ultimate fate of the human species and the world we inhabit. Grappling with Richard Leakey's contention that a massive extinction of the planet's species is well under way, she travels from Madagascar, the 'last living Eden', to the Galapagos, Darwin's natural laboratory, from the Azraq Oasis in Jordan to the Arctic desert of Banks Island, from the rainforests of Suriname to Iceland's volcanic crust. And she shows how, against the odds, individuals have convinced governments to create protected areas, even to transform government policy to seek out new energy sources.

Dancing at the Dead Sea intertwines scientific theory with travel adventure and history. Alanna Mitchell's is a dramatic, and refreshingly optimistic, narrative voice.

'*Dancing at the Dead Sea* is a powerful narrative on the critically important topic of the world's environmental hotspots. This is not a pessimistic tirade, but instead a factual commentary that will convince many, written by a gifted writer with an independent mind. I recommend this book without reservation.'
Richard Leakey

'Captivating . . . easily approachable and digestible while being seriously thought-provoking.'
Chris Stewart

'An important and uplifting read. Alanna marches energetically around the world to find out what every single one of us needs to know: what's really the state of health of our long-suffering little planet? A stirring and ultimately optimistic odyssey.'
Benedict Allen

ISBN 1903 919630

Paperback non-fiction

£8.99